Waterborne & Solvent Based Surface Coating Resins and their Applications

VOLUME V
PART I

The Chemistry and Application of Phenolic Resins or Phenolplasts

VOLUME V
PART I

The Chemistry and Application of Phenolic Resins or Phenolplasts

Dr. T. Burkhart, P. Oberressl, Dr. P. K. T. Oldring

Edited by

Dr. P.K.T. Oldring PhD BA

JOHN WILEY AND SONS

CHICHESTER • NEW YORK • WEINHEIM • TORONTO • SINGAPORE

Published in association with

SITA TECHNOLOGY LIMITED
LONDON UK

Wiley Editorial Offices
John Wiley & Sons Ltd., Baffins Lane,
Chichester, West Sussex, England, PO19 1UD

John Wiley & Sons, Inc., 605 Third Avenue,
New York, NY 10158-0012, USA

VCH Verlagsgesellschaft mvH, Pappelallee 3
D-69469 Weinheim, Germany

Jacaranda Wiley Ltd, G.P.O. Box 859, Brisbane
Queensland 4001, Australia

John Wiley & Sons (Canada) Ltd, 22 Worcester Road,
Rexdale, Ontario M9W 1L1, Canada

John Wiley & Sons (SEA) Pte Ltd, 37 Jalan Pemimpin 05-04,
Block b, Union Industrial Building, Singapore 2057

A catalogue record for this book is available from the British Library

ISBN 0471 978876 1998

Table of Contents

**Chapter III
APPLICATION OF PHENOPLASTS IN COATING SYSTEMS**

CHAPTER I

By P. K. T. Oldring

INTRODUCTION

Chapter I

1 Introduction

Aminoplasts (amino formaldehyde crosslinking agents) and phenolplasts (phenolic resins) are unique in that they are ill defined materials, frequently used at relatively low levels when compared with the major film forming polymers. They are resins in their own right. Their characteristics influence the performance of the film more than any other component.

These resins can crosslink to form chemically resistant, tough, durable films over a wide range of temperatures (typically from about 100°C to 220°C, but even in a few cases, at ambient temperature). The upgrading of the performance of the film of the major film forming resins varies widely, depending upon the type of material selected. The resin which is crosslinked by the amino – or phenolplast is sometimes called the backbone resin and the crosslinking resin the co-cure resin.

Backbone resins are discussed in Pt II § I. In essence they are relatively low molecular weight materials containing functional groups which can react with the functional groups present in the aminoplast or phenolplast. Resins can be divided into thermoplastic or thermoset polymers. The former are typified by having few or no functional groups and they certainly do not undergo any form of chemical reaction when applied in the coating. An example would be the typical emulsion resin (latex) used in traditional interior house emulsion paints. Thermoplastic resins must have high molecular weights (> 50,000 typically) in order to impart their performance characteristics to the films they form. The films are normally formed by solvent evaporation and, as can be expected, do not have outstanding resistance properties, particularly to solvents and chemicals. They can be redissolved or softened after film formation. In contrast, thermoset resins, containing reactive groups, undergo chemical reactions, either with themselves or other materials present in the wet coating film, to form a film with good to outstanding resistance properties. Exposure to solvents should not redissolve the film. Thermoset resins are generally significantly lower in molecular weight than thermoplastic ones. Examples of these thermoset coatings are automotive, heavy duty or can coatings. Chemical reaction is initiated either by heat or by mixing the two components.

Of the many resins which can be used as backbone resins, polyesters, acrylics, epoxies and alkyds are but a few. These resins are the separate subjects for a series of books, of which this is one. The objective is to give a deeper understanding of the chemistry of coatings to those working within the industry whether as development chemists or as raw material and intermediate suppliers.

Both aminoplasts and phenolplasts consist of different categories of materials. In this book, parts I and II, reference will be made to the following;

Phenolplasts

    novolacs

    resoles

Aminoplasts

    urea formaldehyde (UF)

    melamine formaldehyde (MF)

    benzoguanamine formaldehyde

    glycoluril formaldehyde

In all cases, formaldehyde is used during the manufacture of crosslinking agents, although many suppliers of such materials are seriously developing 'formaldehyde free' systems (see below for a discussion on the interpretation of formaldehyde free). As a consequence of their preparation, both aminoplasts and phenolplasts are complex mixtures whose structures are ill-defined. In many cases the reaction mechanisms still have to be fully elucidated. For reactive phenolic resins (resoles) and aminoplasts, there is some degree of self-condensation when they are used as crosslinking agents, and the actual extent of this depends upon many factors, including the nature of both the crosslinking agent and the backbone polymer, the presence and type of catalyst, temperature and other factors still to be determined. In many cases, some degree of self-condensation reactions also occur during the preparation of both aminoplasts and phenolplasts.

Many of today's industrial coatings would not be viable, either technically or commercially, if it were not for the use of phenolplasts or aminoplasts. They offer cost performance benefits, and with the correct selection of crosslinking agent, it is possible disproportionately to up grade the performance of any film forming polymer.

In Volume V Part I phenolplasts are considered. The chemistry of phenolic resins is followed by explanations and examples of their use. It should be borne in mind that the bulk of phenolic resins manufactured are not used for surface coatings. The actual use of phenolic resins in coatings is somewhat limited. Internal lacquers for cans, pails, drums and particularly food cans represent the largest market for the coating sector. Phenolic resins also find application as reaction intermediates in the manufacture of oleoresinous systems. Almost without exception, it is the reactive phenolic resins, known as resoles, which are used in industrial coating systems. Novolacs find a few specialised uses.

In Volume V Part II aminoplasts are considered. One of the major uses of aminoplasts is in the manufacture of compressed boards, such as work tops. As a generalisation, these materials utilise urea formaldehyde resins (UF's). For the more varied, and some would argue more sophisticated, needs of industrial coatings a wider range of crosslinking resins is available. Melamine formaldehyde (MF's) resins were the alternative to UF's. However in more recent times, more sophisticated materials based on benzoguanamine or glycolurils have become available. All four types of product are discussed in this book.

Aminoplasts and phenolplasts are often referred to as cocure resins. Both are film formers in their own right. Phenolic resins can be used as a major film former, but then only for specific end uses (such as internal drum lacquers). Generally films of crosslinking resin on its own are either too brittle, and in the case of many of the aminoplasts, they are also too expensive. Thus for either type of resin it is necessary to use them together with modifying resins. Hence the term co-cure resins rather than crosslinking agents. Aminoplasts are always present at a significantly lower level (typically 5–30% of the total resin solids) than the other film forming resin(s). Hence the terms minor and major film formers. Phenolic resins are typically used at much higher levels than aminoplasts particularly with epoxy resins where 20–40% of the total resin on solids might well in practice be phenolic resin. In some cases it may be higher. Common to the performances of both aminoplasts and phenolplasts in coatings is the general need for the presence of a catalyst, in order to obtain the optimum properties and performance of the cured film.

One of the most difficult concepts to describe accurately is cure. How does one know when a resin is fully cured? What do cure or fully cured mean? Of 100 people asked what is cure, there is a high probability that there will be at least 99 different answers. To some it means the number of MEK double rubs, whilst to others it could mean a particular physical property such as hardness or impact resistance being attained. There is one certainty however, and that is that at full cure, by whatever definition, there will be some degree of residual reactive groups, however small, within the crosslinked film. It is impossible to have no reactive groups remaining as a definition of cure. There will be some reactive groups trapped in the polymer matrix. Thus a pragmatic approach to defining cure will be used here. A film will be considered cured when it meets whatever parameters (such as MEK rubs) required by experience to ensure that it performs as is intended. Definition involving films having less than a given quantity (percentage) of unreacted or low molecular weight material are of little practical use to the coating's formulator. It is a consequence of the laws of science that there will be some residual reactive groups, however small, remaining in the cured film for industrial coatings applied under industrial conditions.

The use of aminoplasts, in particular, has been a major reason for the successful development of a wide range of industrial coatings.

The contributors to this book are, of course, more familiar with their own products than with others. As a consequence many references will be made to the use of products from either Hoechst or Cytec. This in no way implies that these are the only suitable crosslinking agents for a given system. Consult your existing supplier in order to obtain similar products where ever possible.

In order to maximise the benefits of any crosslinking system, it is necessary fully to understand the chemical reactions which occur during the curing process. Nowhere is this more true than in study to understand how aminoplasts and phenolplasts function and the criteria which facilitate the analysis of differences between apparently similar products. As a consequence, it is necessary to cover the chemistry of phenolplast and aminoplast reactions in detail. This has been done particularly for melamine and modified melamine crosslinking agents. The chemistry of the reactions of MF's has been used to illustrate how these materials behave to permit better optimisation of the formulation, rather than give a multitude of formulations without an explanation of the chemistry involved. By necessity understanding and the use of the correct grades of aminoplasts or phenolplasts are heavily dependent upon a thorough appreciation not only of the reaction mechanisms involved, including crosslinking and self-condensation, but also of potentially detrimental processes, such as hydrolysis of the cured film. In addition, a good understanding of the effects of structure, branching and molecular weight of the backbone resin(s), on the properties of the cured film and the type(s) of aminoplast and catalyst required, are essential for any formulator of coatings if maximum performance is to be achieved. To this end there is much discussion in the aminoplast chapters about these topics.

2.   Formaldehyde Free

The interpretation of the term 'formaldehyde free' varies, particularly between the USA and Europe. Some commonly used definitions follow;

- non-detectable formaldehyde in the wet lacquer/coating

- formaldehyde not used in the manufacture of the aminoplast or phenolplast

- formaldehyde not liberated during curing/crosslinking the coating

- formaldehyde not present initially and the system incapable of generating formaldehyde

- non-detectable levels of formaldehyde in both the wet lacquer/coating and non-detectable levels during the curing cycle.

- non-detectable levels of formaldehyde generated during the curing cycle

It goes without saying that everyone involved must understand what others mean when they discuss formaldehyde free systems.

Originally, within our series of four books for Co-curing Resin Systems, we planned to publish a single volume covering both Phenolplast and Aminoplast chemistry, technology and applications.

We have now decided to present this book in two parts, since the areas covered by the two systems are essentially dissimilar.

They relate, of course, on the one hand to Co-curing resin systems for metal and, particularly for metallic packaging, and on the other, for a far wider range of applications throughout the coatings industry.

CHAPTER II

By Dr. T. Burkhart

CHEMISTRY OF PHENOLIC RESINS OR PHENOLPLASTS PHENOL FORMALDEHYDE RESINS (PHENOLICS OR PHENOLPLASTS)

Chapter II

1. History

A. von Bayer[1] in 1872, was the first to report, that phenol reacted readily with formaldehyde under both acidic and alkaline conditions. This kind of polycondensation reaction led to the development of the first synthetic resins and plastics in 1909 by Baekeland[2]. The first reproducible phenolic resins for coatings were prepared under alkaline conditions. But to obtain good coating properties these resins must be baked (i.e. stoved). In order to avoid this limitation, phenolic resins were used together with drying oils. One disadvantage of the former resins was their poor solubility in oils. In 1910 Behrends formulated oil-soluble phenolic resins by polycondensation of phenol, formaldehyde and rosin. These resins were optimized by Albert in 1917, who etherified the reactive phenolic resin, modified with rosin, with glycerol in order to make them soluble in drying oils. But it took a further 18 years, before the first oil-soluble resins of commercial significance were synthesised by Turkington et al. (1935) They were based on p-phenylphenol, whereas most of the oil-soluble resins, which were used in large volume, were based on alkylphenols such as p-tert-butylphenol and p-tert-amylphenol or cyclohexylphenol.

During the next 30 years theoretical and practical work was done by von Euler, Hultzsch, Martin, Megson, Ziegler[3-5] and others, who made it possible to use phenolic resins in many other application areas-beside the ***coating sector*** – as shown in Table 1 overleaf.

As can be seen from Table 1, phenolic resins have many non-coating uses. Industrial coating applications for them only will be considered here. In coatings usage phenolic resins are used mainly in conjunction with other resins, since they participate in chemical crosslinking reactions, to form chemically resistant cured films. Indeed phenolic resins are characterized by the resistance properties which they impart to films containing them. Phenolic resins are highly coloured species due to the existence of many chromophores, so that films containing them tend to be coloured. In many cases this is not a problem. The benefit in film performance often outweighs this fact and in some instances, such as internal food can lacquers, the gold colour imparted by the epoxy phenolic is highly desirable. Some production lines use this coloration for visual confirmation that the can has been properly coated and has undergone a required stoving schedule, during which its gold colour has been generated.

The industrial development of phenolic resins continues despite the long history. As can be seen from Table 1 the importance of phenolic resins is likely to remain considerable. One reason is their manyfold use in industry and another is that raw materials for them can be obtained at reasonable cost from both petroleum and coal. 20% of its market is for coatings and over 85% of phenolic coating resins are used for can coatings, except for varnishes for electrical application.

TABLE 1 : APPLICATIONS FOR PHENOLIC RESINS

Applications for Phenolic Resins:	Examples
Composite wood materials	Particle boards, plywood, compressed laminated wood, fibre boards.
Heat and sound insulation materials	Inorganic fibre materials, inorganic fibres, phenolic resins foam, bonded textile felts.
Industrial laminates and paper impregnation	Electrical laminates, laminated tubes and rods, cotton fabric reinforced laminates, decorative laminates, filters.
Abrasive materials	Grinding wheels, coated abrasives, abrasive papers, abrasive tissues.
Rubber	Vulcanizing resins, reinforcing resins.
Adhesives	Contact adhesives, pressure sensitive adhesives.
Other applications	Foundry resins, photo resists, polymer blends.
Coating Applications for Phenolic Resins	
phenolic modified oleoresinous .... /alkyds	electrical insulating varnishes printing inks spar varnishes
phenolic heat cured	tank-car linings linings for drums and pails coatings for heat exchangers coatings for industrial equipment
phenolic/epoxy heat cured	internal lacquers (can linings) and can coatings

2. Raw Materials

(i) Phenols

Phenols which can be used to produce phenolic resins can be divided into two classes: namely unsubstituted and substituted phenols. Phenol, itself, is the more important. Natural phenol from coal tar accounts for very little of the total world production today. The majority of phenol is synthesized from benzene, a major ingredient from petroleum sources. Many processes exist to produce phenol. A yield of 90% is claimed for the Hook process which also uses benzene as a starting material.

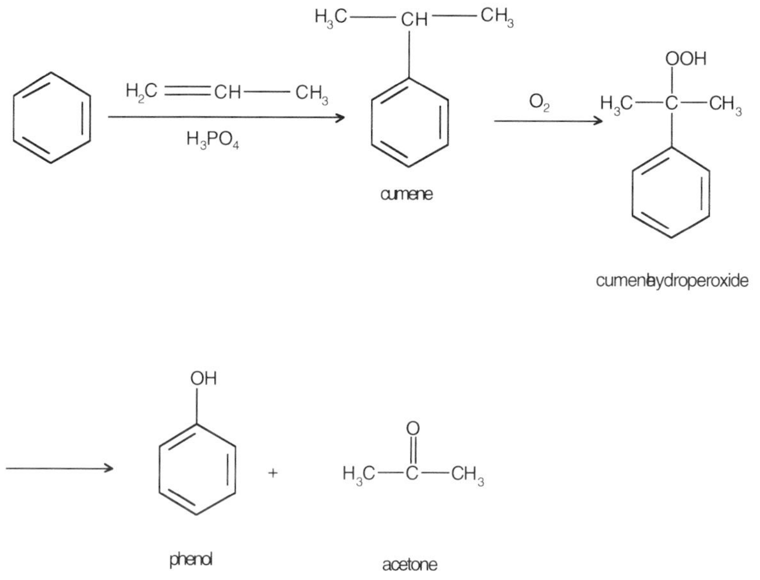

Figure 1 Industrial Production of Phenol.

As the first reaction step, benzene is alkylated in the vapour phase with propylene to give cumene. In the second step cumene is oxidized to cumenhydroperoxide with air, via a free radical mechanism. After that oxidation process, cumenehydroperoxide undergoes cleavage to form phenol and acetone.

(ii) Substituted phenols

(a) *Cresols*
The methylphenols, the so called cresols, are synthesized from either phenol or toluene. There are three isomers which depend upon the relative position of the methyl group in relation to the hydroxyl group as shown in Figure 2.

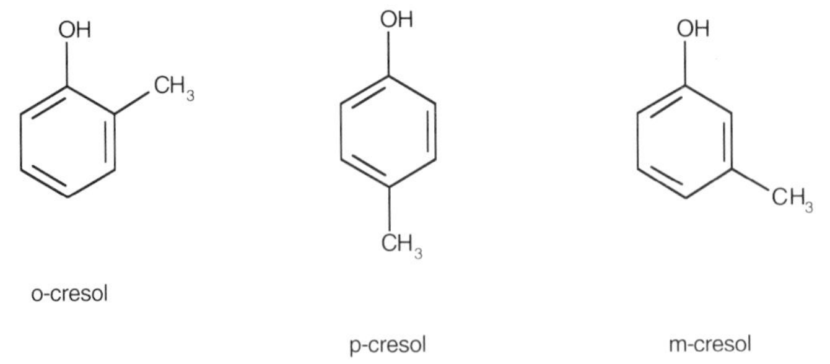

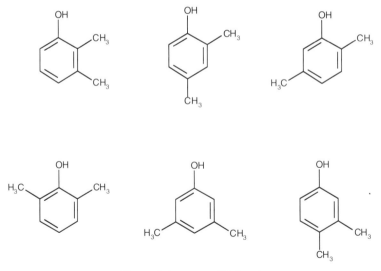

Figure 2 Isomers of Cresol

(b) *Xylenols*

For the preparation of phenolic resins used in can-coatings, the dimethylphenols (the so called xylenols) are very important. There are six isomeric xylenols, all of which are shown in Figure 3. Analogous to phenol and cresols, at one time the six isomeric xylenols were obtained from coal tar and petroleum refining streams. Now xylenols are mainly obtained by the alkylation of phenol with methanol[7]. Xylenols are used to manufacture oil-soluble phenolic resins for surface coatings.

Figure 3 Isomers of Xylenol

(c) *Higher homologues of phenol*

A few synthetic higher homologues of phenol, often called alkylphenols, are used in the manufacture of laminating resins which give products with enhanced flexibility in oil-soluble resins for surface coatings. These alkylphenols are synthesised from phenols or cresols by

Friedel-Craft alkylation with olefins i.e. propene, isobutene or diisobutene[8]. This class of alkylphenols includes 2,2-bis-(4-hydroxyphenylpropane) Bisphenol A (I), p-phenylphenol (II), p-tert-butyl-phenol (III), p-tert-amylphenol (IV), p-tert-octylphenol (V) and p-nonylphenol (VI), although strictly speaking the first two should be classified as aryl phenols.

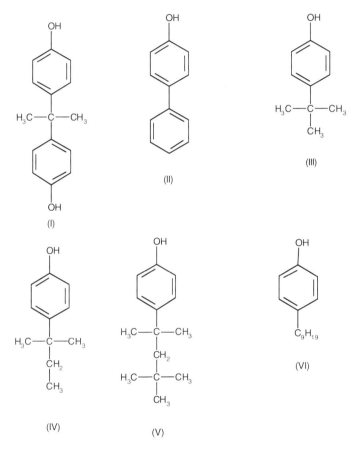

Figure 4 Different Alkyl and Aryl Phenols.

(iii)   Aldehydes

Though many aldehydes can be used to react with phenols to give phenolic resins, only formaldehyde, the lowest molecular weight and most reactive aldehyde, is of major commercial importance. Other aldehydes, for example acetaldehyde, butyraldehyde or furfural are only used to a limited extent. The physical properties of aldehydes are compiled in Table 2[9].

TABLE 2 : TECHNICAL DATA OF DIFFERENT ALDEHYDES

Type	Formula	Melting point [°C]	Boiling point [°C]
Formaldehyde	$CH_2=O$	−92	−21
Acetaldehyde	$CH_3CH=O$	−123	20.8
Butyraldehyde	$CH_3(CH_2)_2CH=O$	−97	74.7
Furfural	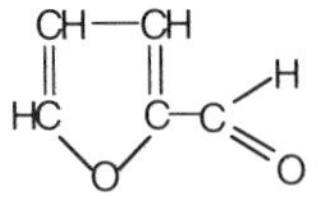	−31	162

(a) Formaldehyde

Since pure formaldehyde is a highly reactive gas, it is commercially used either as a solution in water (known as formalin) or in a solid polymeric flake form, known as paraform or paraformaldehyde. Formaldehyde is synthesized by dehydrogenation of methanol via an oxidation reaction with air[10] as shown in Figure 5.

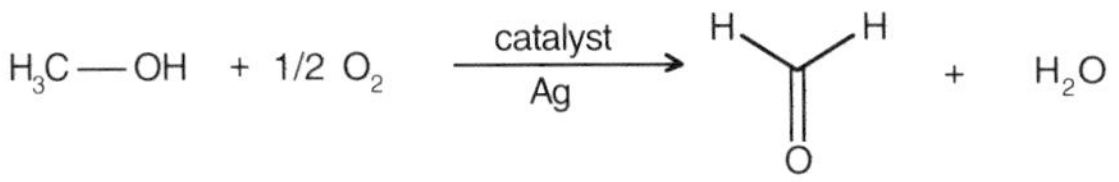

Figure 5 Industrial Production of Formaldehyde

Formaldehyde in aqueous solution does not exist as a pure aldehyde, but rather in hydrated forms such as methylene glycol and low molecular weight glycol ethers as shown in Figure 6[11].

$$CH_2O + H_2O \rightarrow HO - CH_2 - OH + HO - CH_2 - O - CH_2 - OH + \text{etc.}$$

Figure 6 Formation of Glycol Ethers and Methylene Glycol respectively from Formalin

The polycondensation product of formaldehyde is paraform, a white solid, which consists of higher molecular weight polymethylene glycol ethers having a degree of polymerisation of 6–100. These polymers readily unzip (break down) in water, forming monomeric formaldehyde.

(iv) **Phenol-aldehyde polymerisation**

Phenol-aldehyde polymerisation can be divided into two classes, resoles and novolacs, depending upon the pH, the catalysts and the molar ratio of phenol/formaldehyde used.

(a) *Resole resins*

Resole resins are prepared by the interaction of phenol with a molar excess of formaldehyde under alkaline conditions.

The mechanism of the reaction of formaldehyde with phenol under alkaline conditions has been studied by many working groups and is reasonably well understood. In aqueous medium, formaldehyde is present in the form of methylene glycol[11] (see Figures 6,7). The subsequent equilibrum can be proved by different techniques: NMR-, UV-spectroscopy or polarographic methods[12].

$$HO-CH_2-OH \rightleftharpoons \overset{\delta^+}{C}H_2-\overset{\delta^-}{O} + H_2O$$

Figure 7 The Equilibrum of Methylene Glycol in Aqueous Medium

The equilibrum is on the far side of the methylene glycol, but because of the electron withdrawing effect of the hydroxyl group, a positive part charge is located at the carbon atom. Formaldehyde is a so called electrophilic reagent which can therefore react with phenol by a substitution reaction at the para- and/or ortho-positions of the phenol molecule. Consider now, why the electrophilic attack of formaldeyde only takes place at these two positions. Consider the following possible substitution sites shown in Figure 8.

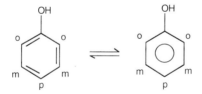

Position	Number
ortho (o)	2
meta (m)	2
para (p)	1

Figure 8 Formula of Phenol and its Reactive Positions

The hydroxyl group of phenol is an inductive electron withdrawing (-I effect) as well as a conjugatively electron releasing group (+ M effect). By the (+) M effect, as can be seen from Figure 9, the delocalisation of an unshared electron pair on the oxygen atom results in increased electron densities at the o- and p-positions.

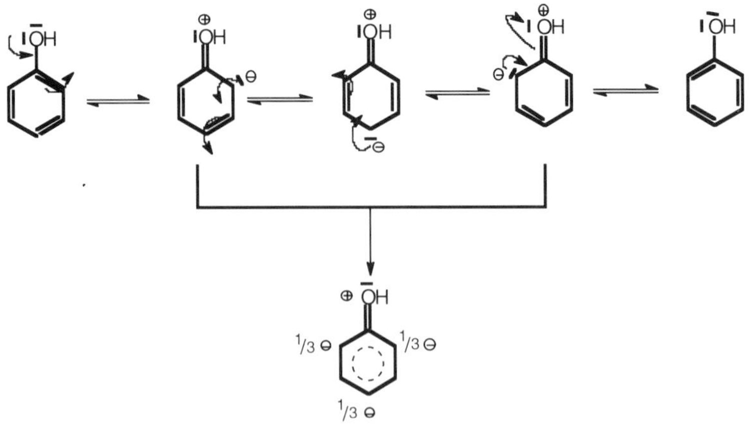

Figure 9 Model for the Delocalisation of a Single Electron Pair of Phenol (⇒ Electrophilic Substitution)

Because of the negative charges on the o– and p–positions, the substitution of phenol by the electrophilic reagent formaldehyde takes place at these two positions. It is known that phenol is more reactive in alkaline conditions than in neutral solution. The reaction of phenol and formaldehyde in alkaline solution therefore results in the formation of o– and p–methylol groups as shown in Figure 10.

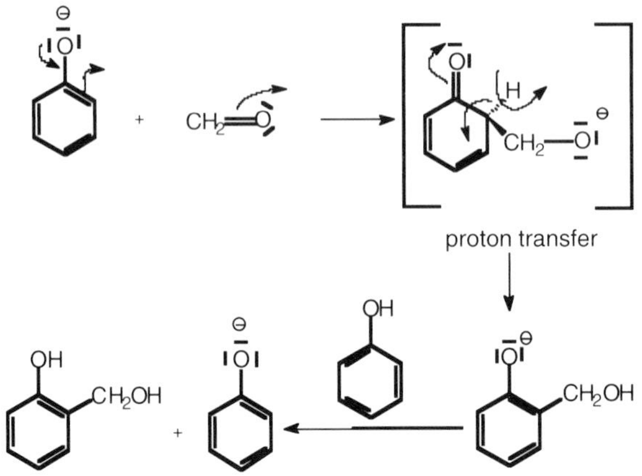

Figure 10 Reaction of Phenol with Formaldehyde under Alkaline conditions.

However it should be noted, that the actual composition of the hydroxyalkylating species has not been established and it is not clear how methylene gylcol would react with the phenoxide ion.

Pople et al.[13] have demonstrated by CNDO/2 calculations (Complete Neglect of Differential Overlap) that the p–position is prefered by an electrophilic attack of formaldehyde in both neutral and alkaline conditions, because of the higher electron density at that position. But the ortho/para substitution ratio is, in addition to electron density, also very dependent upon the type of catalyst which is used.

It is known that the ortho-substitution is considerably favoured if metal hydroxides of the first or second main group of elements are used as catalysts (these metals are listed in Table 3).

TABLE 3 : THE ELEMENTS OF THE FIRST AND SECOND MAIN GROUP

Main group	
first	second
Li+	Mg^{2+}
Na+	Ca^{2+}
K+	Sr^{2+}
	Ba^{2+}

But the ortho-directing effect of the second main group is much larger than that of the first main group. Within a group the ortho-directing effect decreases from the top to the bottom of the Table. That kind of ortho-substitution can be explained by a chelating effect of the elements of the second maingroup, whereas the elements of the first maingroup are not able to build up such a feature (Figure 11).

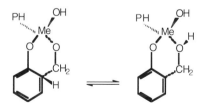

Figure 11 Chelating Effect of the Elements of the Second Main Group

The higher the chelating strength of the cation the greater the ortho substitution. Typical alkalis which can be used to synthesize resoles, which can be used for can-coatings, are listed in Table 4.

TABLE 4 : COMMERCIALY USED ALKALIS TO SYNTHESIZE RESOLES

Alkali	Formula
quaternary ammonium hydroxides	$NR_4{}^+OH^-$
sodium hydroxide	NaOH
monoalkylamine	H_2NR, e.g. R = $-CH_2-CH_3$
dialkylamine	HNR_2, e.g. R = $-CH_2-CH_3$
trialkylamine	NR_3, e.g. R = $-CH_2-CH_3$
ammonia	NH_3
hexamethylentetramine (HEXA)	$N_4(CH_2)_6$

Tertiary amines and quaternary ammonium hydroxides are comparable to NaOH in their activity, whilst mono-, and diethylamine, ammonia or HEXA are essentially weaker catalysts. With alkali, phenol reacts to form the phenoxide ion and then C–alkylation in ortho and para position occurs to form hydroxymethyl substituted phenols. Meta substitution is not detected. From experiments and CNDO/2 calculations it is evident that hydroxymethyl substituted phenols are more reactive than phenol itself, therefore hydroxymethylation continues. The possible products of further hydroxymethylation are shown in Figure 12:

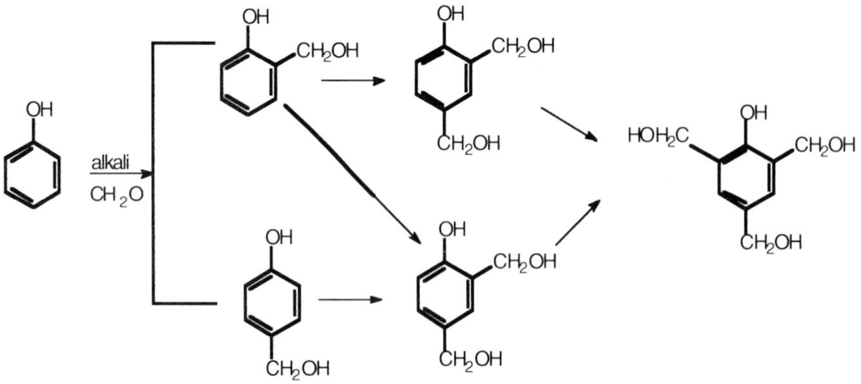

Figure 12 Reaction of Phenol with Aldehyde under alkaline conditions

This is one reason why it is very difficult to synthesize resoles, used as hardeners, for instance, for epoxy resins in the can-coating sector, which are free of mononuclear phenols. Mononuclear phenols are highly toxic compounds (see section 8) so that reaction conditions have to be chosen very carefully to avoid mononuclear phenols in resoles.

In alkali medium the methylolphenols are relatively stable but can undergo self-condensation to form oligonuclear and polynuclear phenols with the formation of methylene bridges by thermal treatment according to the mechanisms shown in Figure 13.

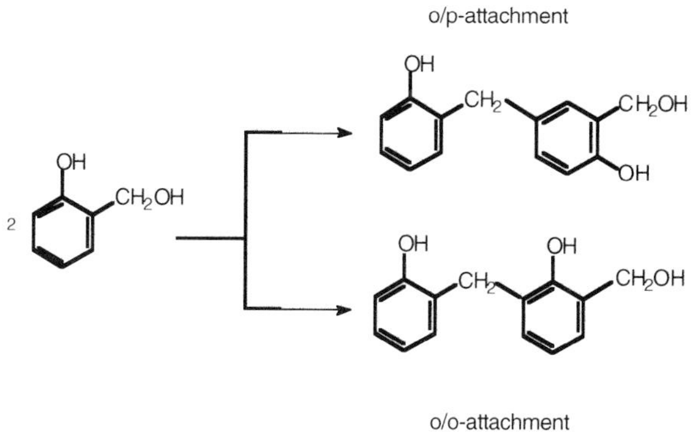

Figure 13 Possible Reaction of Two Methylolated Phenols

These products can arise from two different types of self-condensation. One involves a methylol group (ortho or para) and a hydrogen atom at an unsubstituted o– or p– position in the methylol phenol and phenol respectively as can be seen from Figure 14 and the other one involves two methylol groups, independent of their position (ortho/para).

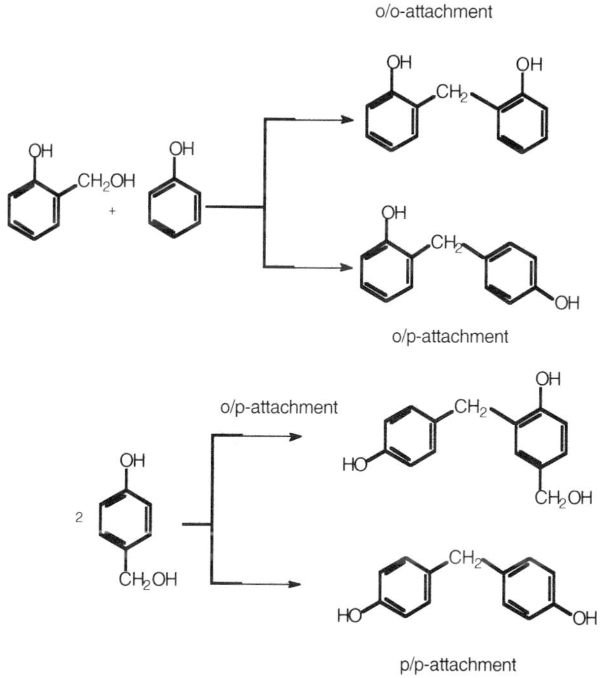

Figure 14 Possible reactions of Methylolated Phenols

It can be shown that the condensation reaction between two methylol substituted phenols proceeds faster than the reaction between hydroxymethyl phenol and phenol[14]. It is also possible for the methylol group to react with another methylol group to form dihydroxydibenzylethers (Figure 15).

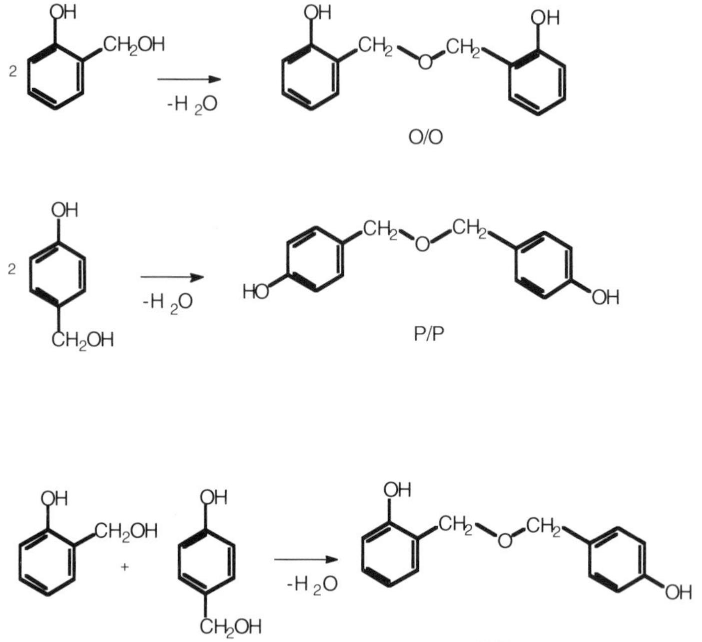

Figure 15 Further possibile reactions of Methylol Substituted Phenols.

The above reactions, which are shown leading to dinuclear phenols, may be repeated so that tri-, oligonuclear phenols are formed, and so forth. Thus the product obtained is a mixture of mono- and polynuclear phenols. The structures of the components of such a mixture are shown in Figure 16.

The pH and temperature determine which type of reaction (formation of methylene bridges or dimethylene ether bridges) occurs. The condensation reaction to form dimethylene ether bridges is the prevalent one under neutral or weak acidic conditions. These bridges are stable up to a temperature of 140°C–150°C. Above that temperature a decay of the ether bridges occurs to form methylene bridges with the loss of formaldehyde as shown in Figure 17.

The formaldehyde liberated is then available for the formation of new hydroxymethyl groups, if further catalyst is present. But under strong alkaline conditions the formation of methylene bridges occurs predominantly. As can be seen from the different schemes.

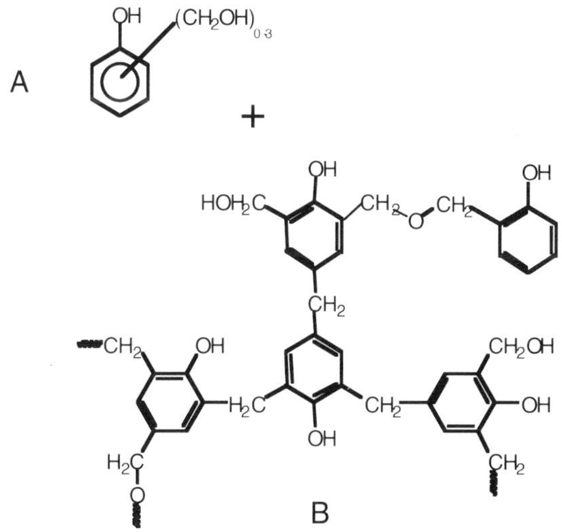

Figure 16 Idealised structures of Phenol Resoles.

A represents the general formula of the methylolated mononuclear species, whereas B represents the idealised structure of the oligo- and polynuclear phenols.

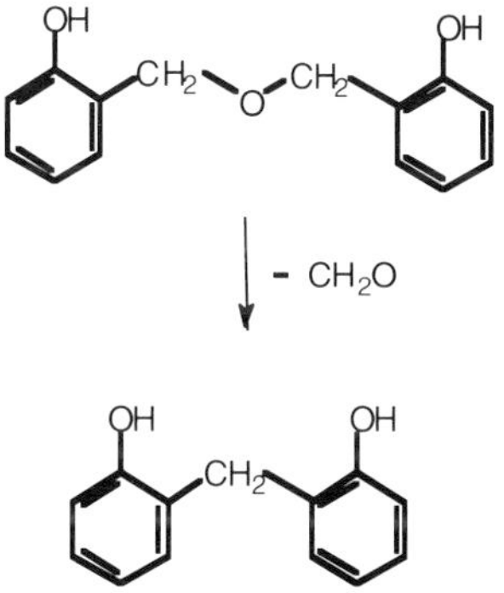

Figure 17 Decomposition of Dimethylenether Bridges at temperatures above 140°C

Once formed hydroxymethyl phenol is able to react in a different manner to form oligo-nuclear polynuclear networks, depending upon the reaction conditions. So the network structure of resoles can be described by two kinds of linkages between the different nuclei, namely;

$-CH_2 - O - CH_2 -$:   dimethylene ether bridge,
$-CH_2-$ methylene bridge,

and three kinds of functional groups at which further reaction can take place:

$-OH$	phenolic hydroxyl group,
$-CH_2OH$	methylol group,
$-CH_2 - O - CH_2-$	dimethylene ether bridge.

The resulting network of the resoles and therefore the resulting functional groups and linkages depends upon different parameters, some of which are listed in Table 5.

TABLE 5 : PARAMETERS WHICH AFFECT THE SYNTHESIS OF RESOLES:

Parameter
reaction temperature and reaction time
modification with other aldehydes and phenols
molar ratio of phenol to formaldehyde
etherification and/or dissolving in organic solvents
type and amount of catalyst
modification with other compounds

To produce resoles for can coatings, different types of phenols are available which can be used (Table 6).

TABLE 6 : DIFFERENT TYPES OF PHENOLS, WHICH WILL BE USED TO PRODUCE RESOLES

Phenols
phenol
cresol (o,m or p)
p-t-butylphenol
p-octylphenol
p-nonylphenol
Bisphenol A
Xylenols

The reactivity of the different phenols towards formaldehyde depends strongly on the reaction conditions (catalyst/solvent) as can be seen from the results of Sprung[15] and Haller/Schmidt[16]. Sprung investigated the relative reaction rates of various phenols with paraformaldehyde and triethanolamine as catalyst, Haller/Schmidt measured the reaction rates in water/methanol with formaldehyde and sodium hydroxide as catalyst. The measured reaction rates are listed in Table 7. It is evident that by using a wide variety of reaction conditions it is possible to produce the desired network structure.

TABLE 7 : RELATIVE REACTION RATES OF VARIOUS PHENOLS WITH FORMALDEHYDE UNDER VARIOUS CONDITIONS

Relative reaction rates of various phenols with formaldehyde	Relative reactivity	
	Sprung	Haller/Schmidt
phenol	1,00	1,00
o-cresol	0,26	1,8
p-cresol	0,35	1,2
m-cresol	2,88	3,1
2,6-xylenol	0,16	2,7
3,5-xylenol	7,74	4,1

3   Resole production

(i)   Introduction

Most of the resoles used for can coating are produced by a batch process. The multitude of resin specifications renders a continuous process uneconomic. The phenolic plant shown in Figure 18 can be used for all the steps involved in batchwise production. Corrosion-resistant or enameled steel is used as the construction material for phenolic resin plants. In the batch process, batch sizes are limited due to the highly exothermic nature of reactions. The heat formation of hydroxymethyl groups is $-20,3$ kJ/mol, whereas it is $-98,7$ kJ/ mol for the formation of methylene bridges[41]. Corresponding to the great variety of phenolic resins, synthesis occurs over a very broad temperature range which extends from 20°C to more than 260°C for phenolic resins modified with natural resins.

Use of aqueous formaldehyde is advantageous from this point of view because water functions as an efficient heat exchanger and thereby reduces any explosion risk. On the other hand, by using paraformaldehyde, production costs can be reduced dramatically as a

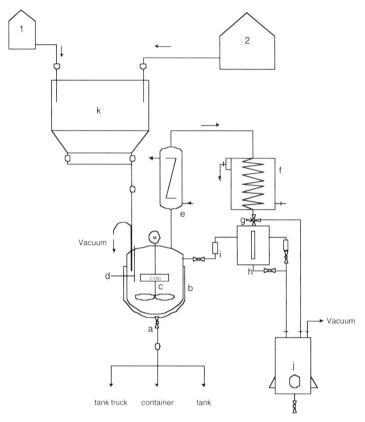

Figure 18 Production Plant of Phenolic Resins

1) Phenol
2) Formaldehyde

a) Outlet tap	g) Three-way tap
b) Heating	h) Reflux and separation vessel
c) Stirrer	i) Reflux return with siphon
d) Thermometer	j) Receiver
e) Evaporator condenser	k) Weigher
f) Main condenser	

result of omission of a distillation process and waste-water (disposal cost). Additionally space and time savings will be increased.

(ii)   **Production of Resoles**

As in the first step phenol and sodium hydroxide are added to the reactor. The mixture is brought to 60°C and aqueous (37% by weight strength) formaldehyde solution is added over a 3h period. After the formaldehyde addition is complete, the reaction proceeds at that temperature during which time almost all the formaldehyde reacts. If the formaldehyde concentration is equal to zero or remains constant, the degree of condensation will be adjusted by isothermal treatment of the reaction mixture at that temperature at which the reaction is controlled.

The degree of condensation is controlled by measurement of the viscosity, if necessary, after dilution with organic solvents, generally butylglycol or methoxypropanol. Depending on the catalyst used, it has to be neutralised with acids and washed out several times with water until the reaction mixture is saltfree. This process will be necessary if the metal ions used are from the first or second main group of the periodic table, for example sodium hydroxide. Most of the acids which will be used for the neutralisation are hydrochloric acid or sulphuric acid. In the case of hexamethylene tetramine, ammonia or amines, whether mono-, secondary-, or tertiary, neutralisation is not necessary because the amines will be incorporated into the network of the resoles by a Mannich-reaction[17] (shown in Figure 19) of the amines with formaldeyde to form hexamethylene tetramine and its intermediates.

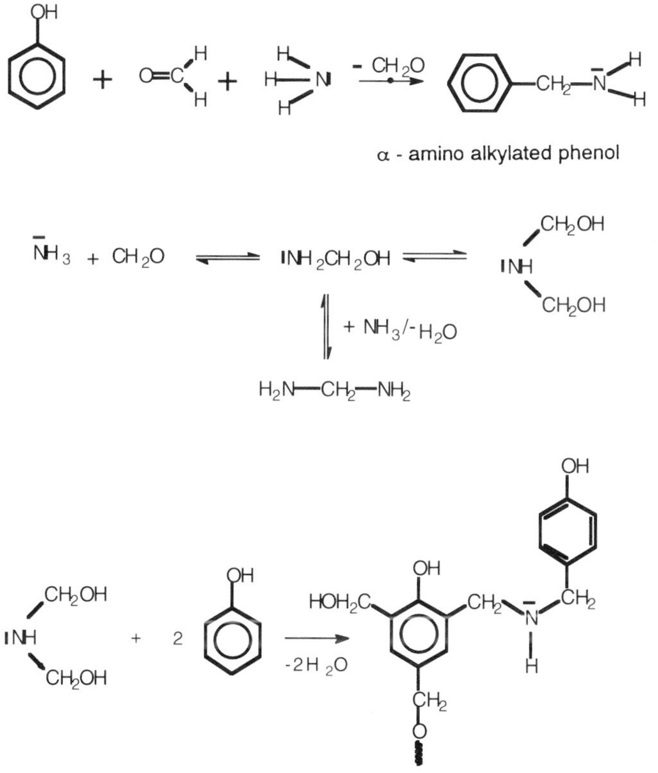

α - amino alkylated phenol

Figure 19 Reaction of Formaldehyde and Amines: Mannich reaction.

Resoles catalysed by amines, where the nitrogen content is greater than 5% by weight, cannot be used in applications where high corrosion resistance or sterilisation resistance is required, for example in can-coating.

After neutralisation and a washing process, the water and water of reaction are distilled off under vacuum at temperatures up to 60°C followed by a subsequent dilution with organic solvents. Alternatively, water can be distilled off azeotropically with an organic solvent as the entrainer e.g. xylene, which is recycled to the reactor. Afterwards, however, the reaction mixture has to be filtered to ensure that it is salt free. Distillation has to be carried out under mild conditions, because during the distillation process the condensation reaction continues and therefore there is a danger of gelation. After distillation the resin will be diluted by polar organic solvents and stored in tanks, vessels or containers.

The gelation depends on different factors, namely

> a)   molar ratio of phenol to formaldehyde
>
> b)   catalyst concentration
>
> c)   temperature

a)   The molar ratio of phenol to formaldehyde can vary widely. The upper limit is determined by functionality. The formaldehyde phenol ratio varies from 3:1 to 1:1 under alkaline conditions. If an excess of phenol is used, a major portion of the phenol cannot react and therefore the resole has to be designated 'poisonous' (free phenol concentration of > 5%) or has to be marked to show a "St. Andrew's Cross" free phenol concentration: 1–5%. Formulations are not considered toxic if the amount of free phenol is below 1%.

If in the case of phenol the molar ratio of formaldehyde to phenol is 3:1, the danger of gelation is very high because of the functionality 3 of phenol. In the case of phenol, therefore, formaldehyde/phenol molar ratios will be chosen between 1.1:1 and 2.2:1.

b)   High alkali concentrations (up to 1 mol per phenolic hydroxyl group) in combination with low temperatures, promote the formation of hydroxymethyl groups which stabilises them. Low catalyst concentrations and high temperatures, particularly, promote condensation reactions and therefore the degree of crosslinking.

c)   The higher the temperature the faster the condensation reactions take place and therefore the degree of crosslinking is very difficult to control.

(iii)   The Laboratory Preparation of a Resole

The reactor used for the condensation reaction and working up of the reaction mixture consists of a 2 l fournecked flask with thermostatically controlled heating, stirrer apparatus, thermometer, feed apparatus and reflux condenser.

Formulation:

1. Charge 592 g of phenol and 292 g of o-cresol to the reactor and adjust pH to 9.15-10.0 by stirring in a concentrated aqueous sodium hydroxide solution.

2. Heat to 60°C and, over two hours add 1095 g of 37% by weight strength of formaldehyde solution.

3. Hold temperature for four hours. Decrease in formaldehyde concentration is monitored by titration with hydroxylamine hydrochloride. At the end of the reaction less than 1% of formaldehyde is present.

4. Neutralise with sulphuric acid adjusting pH to 5-6 at 60°C.

5. Separate salt and wash until salt free.

6. Remove water azeotropically, with xylene so that temperature may reach 60°C max. at 1.5 kPa.

7. Adjust viscosity through organic solvents such as butyl alcohol (butenol) methoxypropanol or butyl glycol.

---

4. Novolac resins

(i) Introduction

Methylolated novolac resins (Novolacs) may be used in the can coating area as a hardener for epoxy resin if excellent adhesion, durability and good impact resistance, independent of the plasticity-high chemical resistance are desirable.

Because of their high chemical resistance, novolacs can be used as an additive for resoles to improve their resistance against chemicals and solvents of the cured epoxy phenolic films.

Novolac resins are normally prepared by the reaction between phenols and formaldehyde under acidic conditions. In order to avoid gelation, novolacs are normally prepared with a molar excess of phenol. In general the formaldehyde/phenol molar ratio is less than 1:1.

(ii) Production

Commonly the reaction is carried out batch-wise in a reactor of the type described in the previous section. The catalysts most frequently used are oxalic acid, sulphuric acid, phosphoric acid and p-toluene sulphonic acid. In practice two alternative processes are available for the preparation of novolacs. They are described over the page.

PROCESS 1: REFLUX PROCESS
1   Feed formaldehyde solution to a reactor containing phenol and acid (usually oxalic acid – 0.3–3% on phenol weight). Stir to maintain a gentle reflux at about 100°C. 2   Continue heating under reflux until formaldehyde is less than 0.1%. 3   Distil water and water of reaction off up to a temperature of 200–220°C. 4   Vacuum distill at 220°C to remove any free phenol. 5   Discharge resin and cool to give solid with a melting point in the range 50–140°C depending upon the molar ratio of phenol to formaldehyde.
PROCESS 2:
Formaldehyde solution will be added dropwise to the reactor, containing phenol and a strong acid under stirring at about 120 to 140 °C with azeotropic distillation of water[18]. For azeotropic distillation an organic solvent (e.g. xylene) will be used as entrainer which is recycled to the reactor. The further processing is the same as the reflux process (Process 1).

Irrespective of the process, the reaction between phenol and formaldehyde under acidic conditions proceeds through a mechanism different to that of the base-catalysed reaction (resoles). The initial step involves the protonation of formaldehyde (methylene glycol) to result in a carbenium ion as shown in Figure 20:

$$HO-CH_2-OH \; + \; H^+ \; \rightleftharpoons \; \overset{\oplus}{C}H_2-OH \; + \; H_2O$$

Figure 20 Protonation of Formaldehyde

The phenol then undergoes electrophilic substitution with the formation of o– and p–methylol groups as shown in Figure 21.

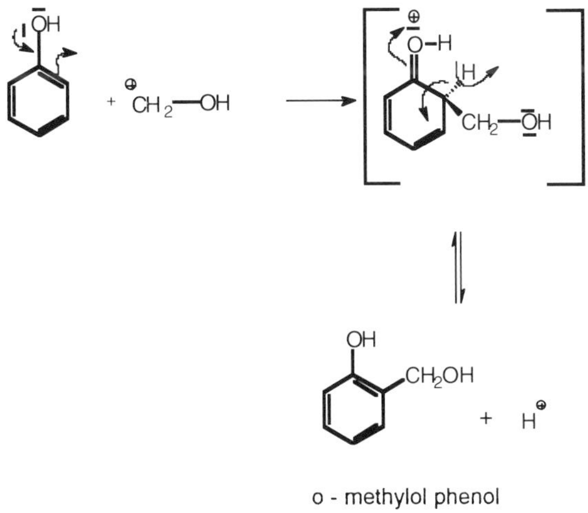

o - methylol phenol

Figure 21 Reaction of the Carbenium ion with Phenol

In the presence of acid, the initial products, o– and p–methylol phenols, are only present transiently in very small concentrations. They react with free phenol to form dihydroxy-diphenylmethanes as shown in Figure 22.

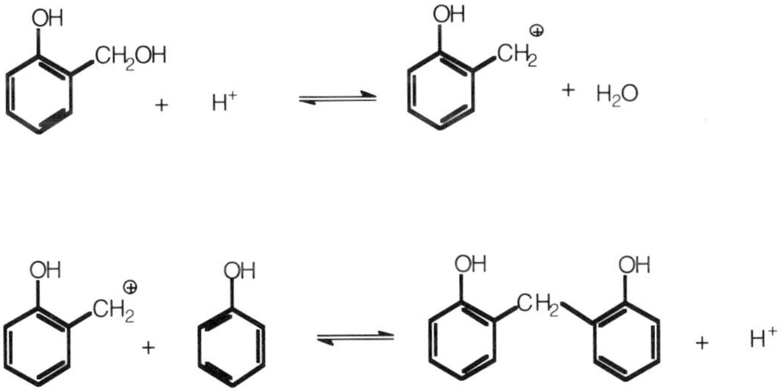

Figure 22 Reaction of Methylolated Phenols with further Catalyst

Once formed, o– and/or p–methylol phenol are able to build up isomers of dihydroxy-diphenylmethane by further reactions with phenol as shown in Figure 23.

During further condensation reactions, the concentration of dihydroxydiphenylmethanes subsequently falls, as the concentration of polynuclear phenols increases, by further methylolation and methylene linkage formation as shown in Figure 24.

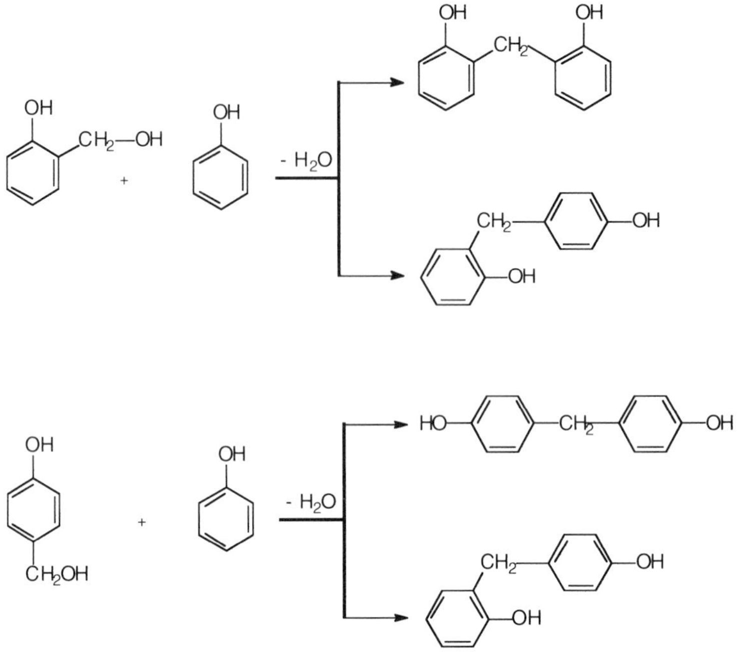

Figure 23 Reaction of Methylolated Phenol with further Phenol

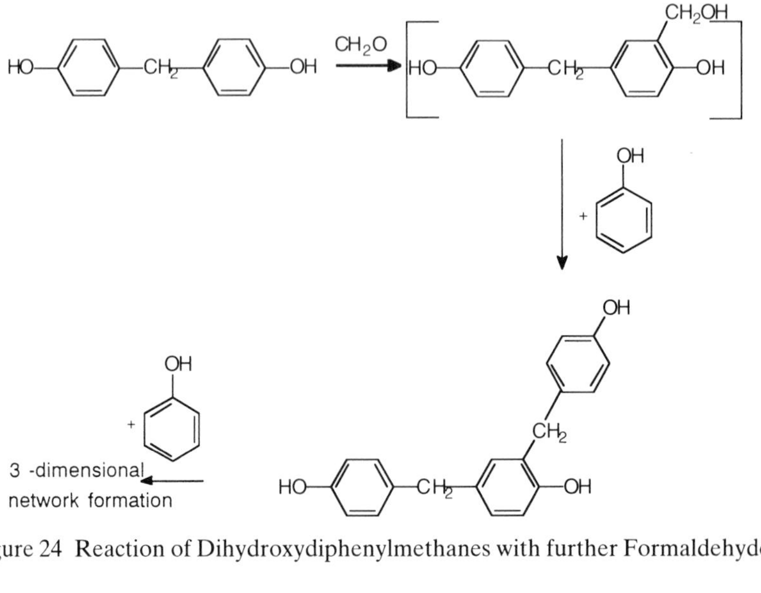

Figure 24 Reaction of Dihydroxydiphenylmethanes with further Formaldehyde.

Reactions of the type above continue until all the formaldehyde had been consumed. The final product therefore consists of a complex mixture of polynuclear phenols linked by o– and p–methylene groups. A typical novolac chain might be represented as shown in Figure 25.

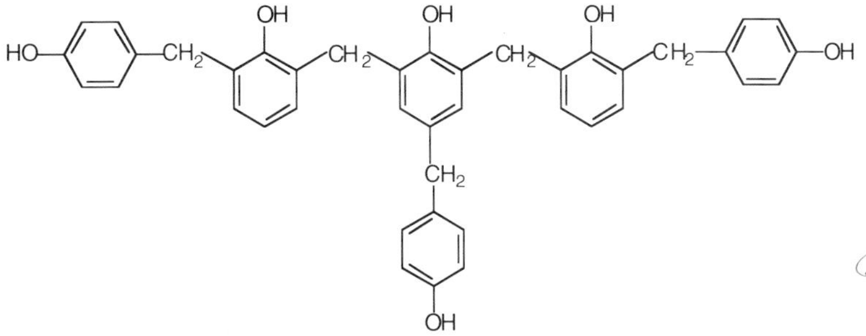

Figure 25 Idealised Novolac structure.

The novolac consists of phenol linked together via methylene bridges. In contrast to resoles they contain no reactive methylol groups or methylene ether bridges respectively and therefore they are not capable of crosslinking on heating by themselves. In order to convert novolacs into network polymers the addition of an auxiliary chemical crosslinking agent is necessary, for example a methylene donor such as hexamethylene tetramine.

(iii)   The Laboratory preparation of a Phenol Novolac by Azeotropic Distillation

The reactor used for the condensation reaction and working up of the reaction mixture consists of a 2 l fournecked flask with thermostatically controlled heating, stirrer apparatus, thermometer, feed apparatus and reflux condenser with a water separator. The water separator has a siphon of adjustable height, which enables water to be separated off continously.

FORMULATION:

1. Charge 940 g of phenol, 40 g of xylene and 40 ml of 0.5 N sulphuric acid to the reactor and the water separator is filled with xylene. Heat to about 140°C. Water introduced with the catalyst distills off.

2. Add 616 g of a 37 % by weight strength formaldehyde solution, over three hours at 140°C. The separation of water starts as soon as the formaldehyde is metered in and ends five minutes after the formaldehye addition has stopped.

3. Add 20 ml of 0.5 N sodium hydroxide. Water introduced with the sodium hydroxide is removed by a recycle distillation up to a temperature of 170°C. A total of 608 g of aqueous phase is produced with a formaldehyde content of 0.15% by weight.

4. Replace the reflux condenser and water separator with a receiver and distil off volatiles, under normal pressure to 210°C and then under vacuum of 60 m bar for one hour at 210°C. In the reactor flask there will be 913 g of a phenol novolac, with a melting point of 92°C and a viscosity of 2790 mPa.s (1:1 in methoxypropanol measured at 20°C).

This type of phenol novolac can be used as an additive for resoles to cure epoxy resins of 7 and 4 type. The chemical resistance of the resulting film, cured at 200°C for 12 min. will be very high in comparison with commercially used resoles. To increase the plasticity, it is preferable to use co-condensation products e.g. cresole-phenol novolacs.

5. Surface coating resins

Whilst phenolic resins in surface coating applications can be and, for some applications are, used on their own or as the predominate film forming resin, they are normally used in conjunction with other resins. Normally phenolic resins are used with epoxy resins where the phenolic resin acts as a crosslinking resin for the epoxy. The ratios of epoxy to phenolic vary, depending upon end use, but generally more epoxy is used than phenolic

(i) **Introduction: Properties of Phenolic Resins used as Curing Agents for Epoxy Resins**

Phenolic resins have many advantages, some of which are listed below[19]:

- High adhesion to many surfaces
  for example, tin plate and aluminium

- Low water vapour and oxygen diffusion

- Excellent resistance to chemicals
  for example, mineral and organic acids

- Excellent resistance to heat
  short periods – up to 30 min. – of dry heat
  exposure to temperatures as high as 315-370°C

Because of their excellent qualities, and food contact approvals, phenolic resins are the major co-curing resins for epoxy resins in can-coating internal lacquers.

The use of straight resole and novolak resins derived from phenol itself in surface coatings, as a hardener for epoxy resins, are very limited, mainly because of their incompatibility with '7' and '4' types, the most commonly used epoxy resins in can-coatings. A further major drawback is the brittleness of the cured films, which is a result of the 3-dimensional network of the cured film, is unacceptable with these types of phenolic resins. Two possible ways of using phenolic resins in coatings by making them more flexible are considered. Both routes can be used to meet all of today's coating requirements.

(ii) **Internal and external flexibilisation of phenolic resins**

Internal flexibilisation

The flexibilisation of phenolic resins by chemical reaction can be classified as internal flexibilisation. This can be achieved by any of the following;

- Modification with other compounds, such as rosin, polyvinylbutyral, polymeric plasticisers of the polyurethane type, for example FU 112®, a product from Vianova Resins GmbH.

- Use of substituted phenols such as cresols or alkyl phenols

- Etherification of the hydroxymethyl groups

- Etherification of the phenolic hydroxyl groups

External flexibilisation

Flexibilisation of phenolic resins by physically mixing with suitable plasticisers would be classed as external flexibilisation. One method of flexibilising phenolic resins, produced only from phenol itself, is to compound them with an organic plasticiser. The flexibilisers which can be used are listed in Table 8:

TABLE 8 : FLEXIBILISERS WHICH CAN BE USED WITH PHENOLIC RESINS

Polyvinylbutyral
Polyvinylformal
Alkyds
Phenoxy resins
Polyamides
Polyurethanes

But this method has two main disadvantages:

- The price of the flexibiliser is, on average, more expensive than that of the phenolic resin itself.

- There is an additional mixing step at the end user including additional expenditure due to the involvement of technical personnel, specialised equipment and time

(iii)   **Compatibility of Phenolic Resins with Epoxy Resins**

To ensure maximum flexibility, the combined species must be compatible. In general, compatibility occurs when a reaction takes place between the phenolic resin and the flexibiliser. Conventional flexibilisers, such as dibutylphthalate, tricresylphosphate etc., cannot be used due to their incompatibilty with the phenolic resin during baking, when they are able to migrate to the paint surface. The most widely used flexibilisers, today, in can-coatings, are epoxy resins, although polyvinyl butyral finds limited usage. Nonetheless because of the disadvantages of using external flexibilisers, the producers of phenolic resins endeavour to produce flexible phenolic resins which are compatible with epoxy resins, without the need for additional external flexibilisers.

Phenolic resins synthesised from phenol itself are, in addition to their high degree of crosslinking, for the most part incompatible with epoxy resins. The chemical structures of the polymers which are to be combined play a dominant role in determining compatibility[20-22]. Since the driving force for polymer miscibility comes largely from the favourable heat of mixing, it may be worthwhile to re-examine the familiar expression, $\Delta \varepsilon$, for the change in pair contact energy

$$\Delta\varepsilon = \tfrac{1}{2}\,(\varepsilon_{11} + \varepsilon_{22}) - \varepsilon_{12}$$

where the ε's are the energies of contact dissociation between like and unlike species. The concepts of specific interaction and complementary dissimilarity focus on the term ε_{12}. It is obvious from the above equation that an appropriate change in ε_{11} or ε_{22} could lead to a favourable value of $\Delta\varepsilon$. In other words, there are two ways of promoting compatibility, namely:

- increasing interaction between unlike species, for example hydrogen bonding between phenolic resins (phenolic hydroxyl group, methylol groups or dimethylene ether bridges) and ether groups, hydroxyl groups respectively of epoxy resins.

- decreasing interactions between like species, such as decreased hydrogen bonding between phenolic resins.

To increase the compatibility between the phenolic and epoxy resins it is better to use resoles, because they are methylolated. They increase the possibilities for hydrogen bonding with the epoxy resins. To increase the flexibility of the resulting epoxy-phenolic network the phenolic resins (resoles or novolacs) are often co-condensation products of phenol and cresol or alkylphenol

(iv)   Resoles and Novolacs: Co-condensation products of different Phenols

The most commonly used reaction route to synthesize internally flexibilised phenolic resins incorporates different homologues of phenol. The ones most frequently used are listed in Table 9.

Experiments have shown that the combination of homologues of phenol with functionality's of 3 and 2 in a ratio 7:3 to 3:7 by weight, give the internal flexibility and compatibility with type '4' and '7' epoxy resins neccessary to meet the requirements of today's can-coatings. To synthesize resoles consisting of phenol itself and higher homologues of phenol (for example, cresols or alkylphenols), a special kind of reaction technique has to be used. After the commercial resole synthesis (§3 iii), a reaction between m-cresol and phenol cannot take

TABLE 9 : THE HOMOLOGUES OF PHENOL AND THEIR FUNCTIONALITY

Homologues of phenol	Functionality
phenol	3
o-cresol	2
m-cresol	3
p-cresol	2
p-t-butylphenol	2
p-octylphenol	2
bisphenol A	4

place, because of the different reaction rates of phenol and m-cresol with formaldehyde. The reaction rate of m-cresol with formaldehyde (Table 7) is about three times higher than that of phenol. Therefore during the commercial resole synthesis, the m-cresol readily reacts with formaldehyde to form a polynuclear network whereas the major part of the phenol remains unreacted. It is possible that this can lead to phase separation during resole synthesis. To avoid this kind of reaction two possibilities exist to form a network between phenol and homologues of phenol, characterised by large differences in reaction rates towards formaldehyde

The first one is that the homologues of phenol – within the mixture in which the resole has to be prepared – with the lowest rate of reaction with formaldehyde should be prereacted with formaldehyde under alkaline conditions to form highly methylolated mononuclear compounds. This kind of compound can then react in a second step with the homologues of phenol with the highest rate of reaction with formaldehyde. By this technique a homogeneous network can be built up, but this type of process is very expensive since it is multistaged.

In the 80's another technique to build up a homogeneous network of co-condensation products of homologues of phenol with different rates of reaction with formaldehyde was developed by Hesse et al.[18]. Within this reaction procedure, the different rates of reaction of the phenols with formaldehyde were graded and offset by using high temperatures, in contrast to commercial resole synthesis. It is a special kind of synthesis to produce novolacs.

(a) *Manufacture of Phenol Homologues:*

Homologues of phenol are added to the reactor, together with xylene as entrainer, and heated up to 100°C. At this temperature under stirring a strong acid (organic or inorganic) is added. Formaldehyde solution is added dropwise to the reactor at about 120–140°C with azeotropic distillation of water and water of reaction. Xylene as entrainer is recycled to the reactor. The further working up is similar to the reflux process.

The different phenols of the resulting novolac are linked via methylene bridges, so that the only functional groups which are able to react with the epoxy system are the phenolic hydroxyl groups. Because of this, the rate of reaction between the novolac and epoxy system is, compared to resoles, decreased. The rate of reaction can be increased by further methylolation of the novolac, via resole synthesis, or by a greater amount of catalyst, such as phosphoric acid.

(b)   *Etherification of the Hydroxy Methyl Groups of Resoles*

Another approach for meeting the requirements of flexibility and compatibilty is the etherification reaction of methylolated phenolic resins-resoles as well as methylolated novolacs – with different alcohols for example, butanol[23], as shown in Figure 26.

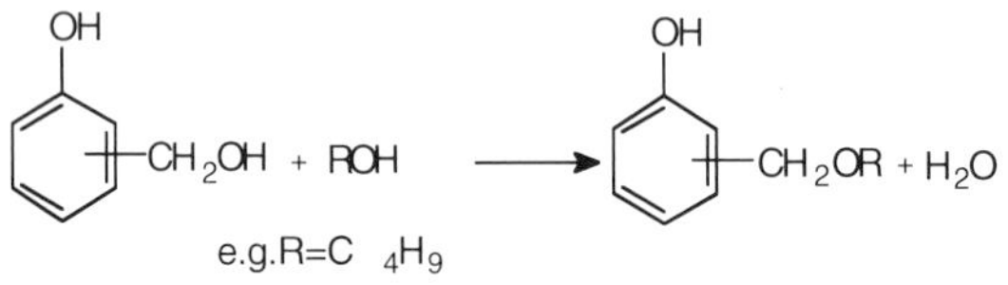

Figure 26 Etherification of the Methylol Groups of Resoles

To etherify the hydroxymethyl groups, the resole, diluted in an organic solvent as entrainer, is heated together with alcohol, for example, butanol. The water formed is azeotropically distilled off with organic solvent (for example, xylene) which is recycled to the reactor. The etherification reaction is controlled by the measurement of the solubility in xylene or cyclohexane.

Etherification on the Laboratory Scale

(c) Formulation – Preparation of Methylol Resoles by Etherification

1. Charge 340 g of a phenol resole (produced from Formulation 3 iii) to a three necked flask, fitted with a Dean Stark head.

2. Fill the head with butanol and xylene (Ratio 1:1). Add 570 g of butanol and 329 of xylene. Stir at 50°C until all components have dissolved.

3. Adjust pH to 6, with phosphoric acid (50% in butanol) under vigorous stirring.

4. Heat to 130°C. Separate distilled off reaction water and butanol from the Dean and Stark head and recycle the butanol. Samples of the resole solution with xylene become cloudy. After 18 ml of water have been distilled off a sample can be diluted at 23°C with 5 parts of xylene and will remain clear. The higher the compatibility with xylene, the higher the degree of etherification.

Etherified resoles exhibit a higher solubility in organic oxygen containing solvents (for example, butanol, methoxypropanol or butylglycol) and show improved flexibility. Their reactivity, however, is reduced. Additionally polyhydroxy compounds are used to plasticise resoles, (for example, glycol, glycerine, hydroxypolyesters and polyvinylbutyral) as shown in Figure 27.

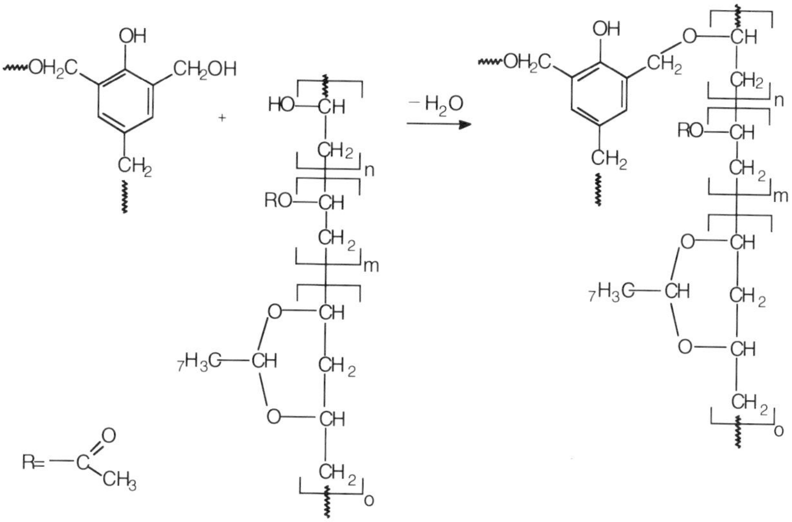

Figure 27 Etherification of a Resole with Polyvinylbutyral

Another reaction mechanism leading to improved flexibility is the etherification of the phenolic hydroxyl group. The reactivity of phenolethers with formaldehyde is drastically reduced compared with phenols. This means that the resole is prepared in the first stage. In the second the phenolic hydroxyl group is etherified with strong electrophiles, such as halogen containing organic compounds, (for example, allylchloride, alkylbromide or epichlorohydrin) in the presence of alkali.

(d)   *Etherification of the Phenolic hydroxyl group (o-alkylation)*

O-alkylated resoles or novolacs can be prepared with strong electrophiles (for example, alkyl bromide, allyl chloride, epichlorhydrin and dialkyl sulphates) in the presence of sodium hydroxide, via a Williamson synthesis[24] as shown in Figure 28.

This especially applies especially to novolacs. Resoles o-alkylated with allyl chloride have been used over several years in conjunction with epoxy resins on account of their very good compatibility, flexibility and enhanced drying properties.

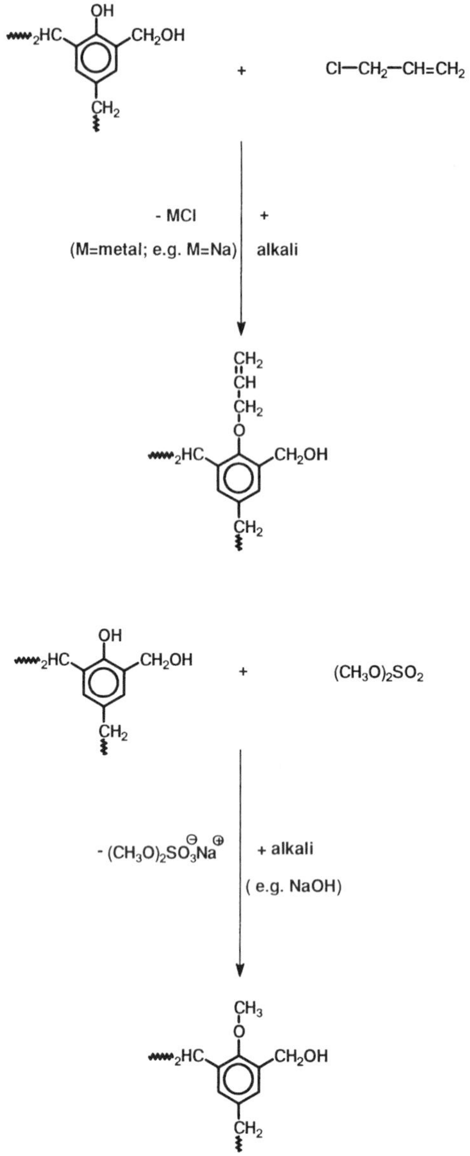

Figure 28 O-alkylation of Novolacs or Resoles via a Williamson synthesis.

(e) *Properties of etherifed novolacs or resoles*

The properties of etherified resoles are listed in Table 10:

TABLE 10 : PROPERTIES OF ALKYLATED PHENOLIC RESINS

c-alkylated/o-alkylated
higher solubility in organic solvents
improved flexibilty
reduced reactivity

In comparison with c-alkylated resoles, the o-alkylated resoles have improved alkali resistance, light fastness and thermo-oxidative stability. But the o-alkylated respectively c-alkylated resoles have the important disadvantage, of reduced reactivity. In fact, reactivity, measured by solvent resistance using a methyl-ethylketone (MEK) "double rub" technique, is very low in comparison with non alkylated alkylphenol/phenol-resoles.

To avoid reduced reactivity, the flexibility can be achieved by combinations of alkyl substituted phenols and phenol, itself, to produce resoles

(f) *Modification with rosin*

Resoles or novolacs modified with rosin[25] represent a specialised type of plasticised phenolic resin which can be used as an additive for resoles and therefore as a co-curing resin for epoxy resin.

As a natural resin, rosin consists of different isomeric types of rosin acid. One of them, abietic acid, is able to react with a resole, via its unsaturated centre. From Figure 29 it is evident that a Diels-Alder reaction has taken place, via an ortho-quinone-methide structure of the resole.

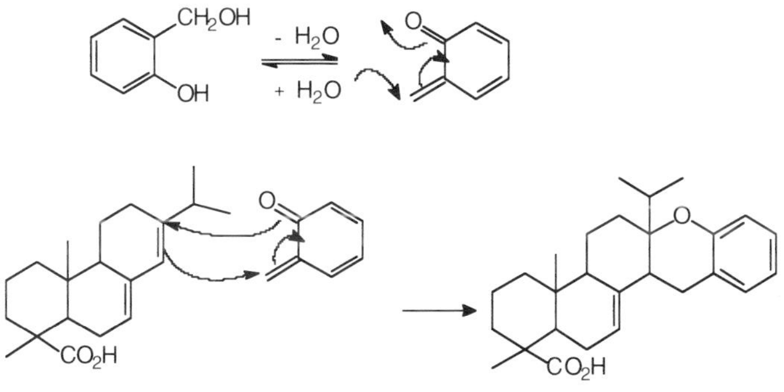

Figure 29 Reaction of rosin with resole via Diels-Alder

The other reaction mechanism which is discussed is the CH-abstraction of an allyl-hydrogen atom (see Figure 30).

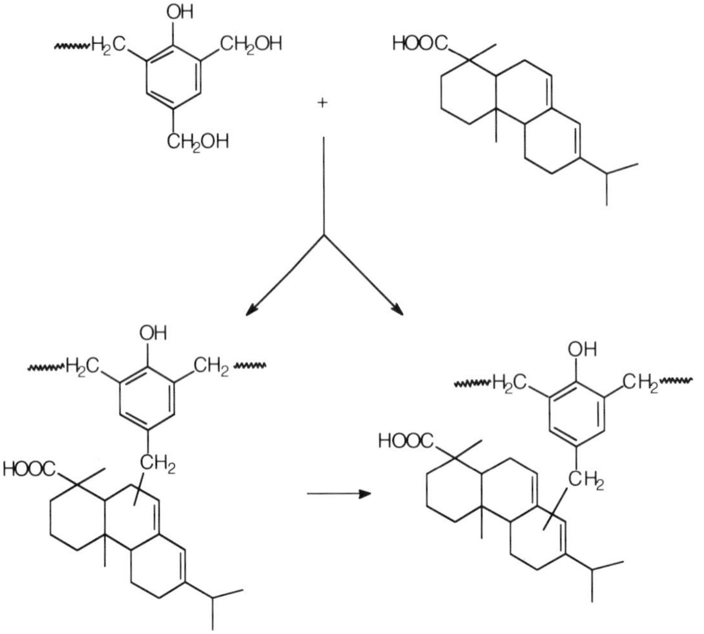

Figure 30 Reaction of Rosin with Resole via CH-abstraction

Reaction products from rosin and resoles are known as albertol acids. The resoles and novolacs prepared from phenols and rosin, showed an improved chemical resistance compared with commercial resoles. A further advantage of such products is high compatibility with epoxy resins. On the other hand a disadvantage is low reactivity towards epoxy resins. Only rosin modified resoles can be used, as opposed to novolacs, as an additive for resoles

6.   Hardening reactions of epoxy-phenol (resoles/novolacs) systems

The degree of crosslinking of epoxy phenolic (resoles/novolacs) systems determines the performance characteristics of cured films using them. By using '4' and '7' type epoxies, there is a low concentration of epoxy groups. Therefore the classical reaction between the phenolic hydroxyl group and the epoxy group, known from the literature[26], is very limited. For that reason other reaction possibilities have to occur to enable a three dimensional network to be built up. Figure 31 presents a schematic representation of some possible reaction mechanisms. Linear addition of a phenolic hydroxyl group with an epoxy group proceeds by the path designated as k_1 (Figure 31) to produce an alcohol group pendant to the chain backbone (secondary hydroxyl group). The same product results from the reaction path designated as

k_2, by the reaction of the hydroxy methyl group (methylol) of the resole with an epoxy group. The successful reaction between o-methoxy-2,6-dimethylol-p-cresol and isobutyl glycidyl ether has been described[37] and this proves the above reaction path shown as k_2. Subsequent to the formation of this species, the terminal epoxide group can react with this pendant hydroxyl group to form a branch, illustrated as reaction path k_3. This kind of crosslinking can also take place through the reaction of the hydroxy methyl group of resoles with the pendant hydroxy group of the epoxide, via an etherification reaction, shown as path k_4. Continuation of these branching reactions will ultimately produce a crosslinked material

Besides the possibilty of forming a homogeneous three-dimensional network, via chemical reactions, there is also the possibility and probability that the epoxy phenolic system builds up an interpenetrated network (IPN) due to its compatibility[27].

(a) *Catalysis of the epoxy-resole, novolac reactions*

In general, the phenolic system used as a hardener (co-cure resin) for '4' and '7' type epoxy systems for can coatings, will be stoved at 185-205°C for 10-12 minutes. At this temperature, the non-catalysed reactions for the epoxy-phenolic system become important due to their reactivity, determined by using "MEK double rub" techniques. The crosslinking of the resulting non-catalysed epoxy phenolic network is very low so that solvent resistance is also

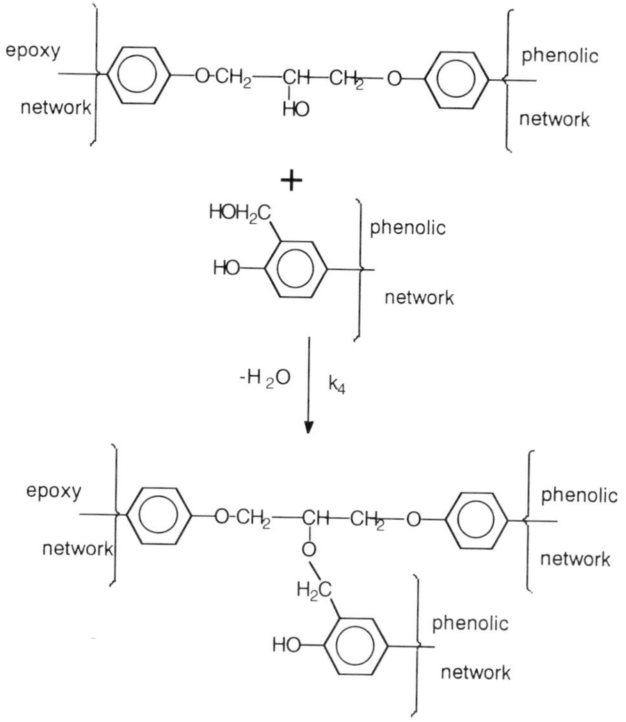

Figure 31   Schematic representation of the reaction of Phenolic Resins (resoles/novolacs) and an Epoxy Resin based on Bisphenol A.
Linear addition denoted by k_1 and k_2;
Branch formation denoted by k_3 and k_4

low in comparison with that which results from a catalysed reaction. The uncatalysed epoxy phenolic reaction has been intensively investigated by Shechter and Wynstra[28]. They found that the phenol, itself, and therefore the phenolic system (resole) catalyses the epoxy phenolic reaction, especially the secondary hydroxyl epoxide reaction (see Figure 32).

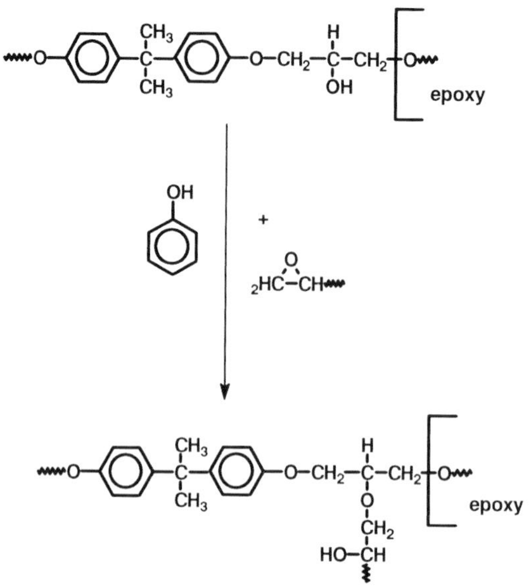

Figure 32 The Phenol catalysed reaction of the secondary Hydroxyl-epoxide reaction.

Other working groups[29] have demonstrated, via kinetic data, that phenol would catalyse the ring-opening reaction of the epoxide as depicted in Figure 33.

To improve solvent resistance, it is necessary to build up a higher crosslinked network. Therefore an external catalyst will be needed to promote the epoxy phenolic reaction. Phosphoric acid is the most commercially used catalyst for epoxy phenolic reactions in can coatings. A commercially available example of this catalyst is Additol XK 406®.

The epoxy phenolic reaction occurs as follows. In the first stage, the reaction takes place very readily due to the low viscosity and therefore the high mobility of the reactants. With increasing crosslinking, however, the mobility of the reactants decreases. To build up a 3 dimensional network it is important that the polymerization mechanism shows no diffusion limitations. The most plausible mechanism for describing the so called bulk-state epoxy phenol reaction is that which involves a transfer of charge or proton between the reactants. The catalyst used should, therefore, be able to promote an ionic polymerisation mechanism. This kind of mechanism would show diffusion limitation only very late in the cure, because the charges are highly mobile. This type of mechanism is also preferred, because the reaction requires coordination between only two molecules at a time, rather than three or more molecules, where the possibility of contact with each other is reduced.

The phosphoric acid catalysed mechanism can, therefore, be explained in the following way. From MNDO calculations (Modified Neglect of Diatomic Differential Overlap) it is known that the electron density of the hydroxy methyl group (methylol) in resoles is higher than

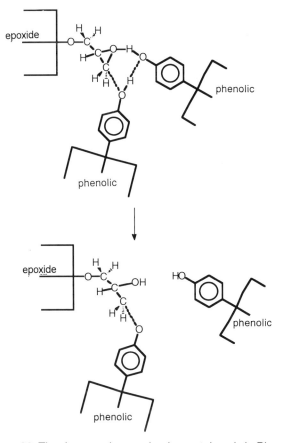

Figure 33 The ring-opening mechanism catalysed via Phenol

that of the secondary hydroxyl group of the epoxide or the phenolic hydroxyl group. Therefore, there is a very high probability that the protonation of the hydroxy methyl group will be greater than that of the secondary hydroxyl group of the epoxide, as shown in Figure 34

From Figure 34 it is further evident that the resulting carbocation is resonance stabilised. The mechanism depicted in Figure 34 confirms this. The structure which results from this resonance stabilisation is the so called 'quinone methide' structure. Furthermore, either the carbocation or the quinone methide structures are able to react with the secondary hydroxyl group of the epoxide, either through an electrophilic attack of the carbocation at the secondary hydroxyl group of the epoxide or by an addition reaction of the secondary hydroxyl group with the double bond of the quinone methide structure. The different mechanisms are shown in Figure 35/36.

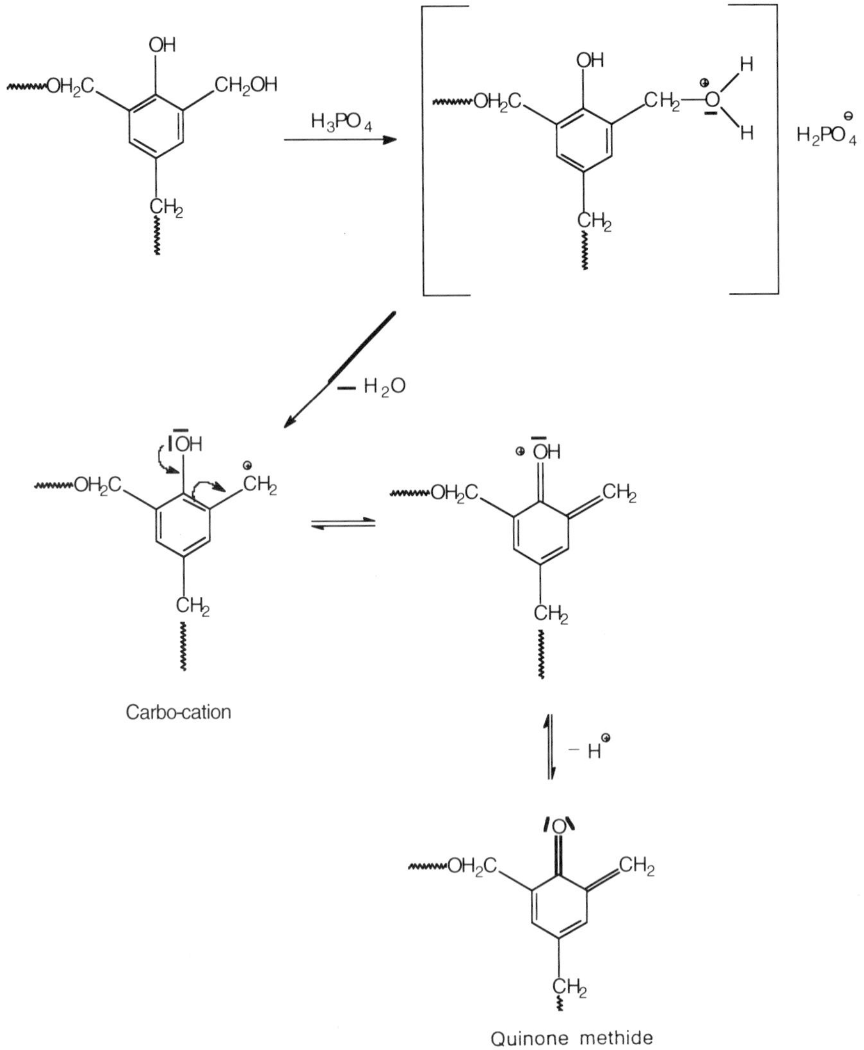

Figure 34 Protonation of the Hydroxy Methyl Group, via a Phosphoric Acid catalyst.

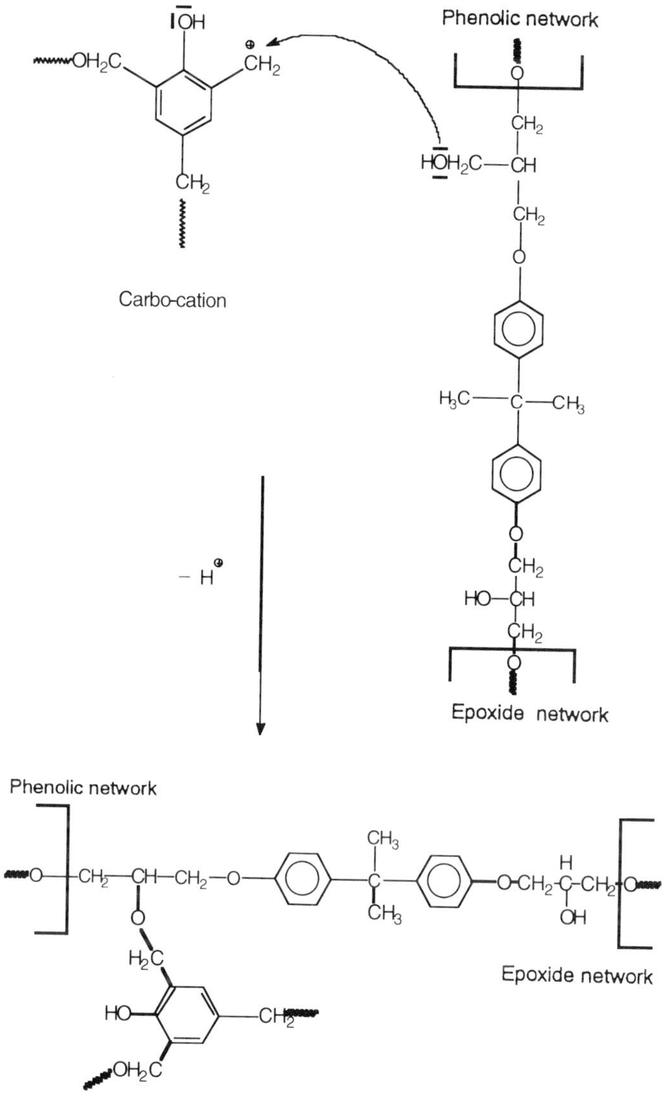

Figure 35 Reaction of the carbocation with the secondary Hydroxyl Group of the Epoxide

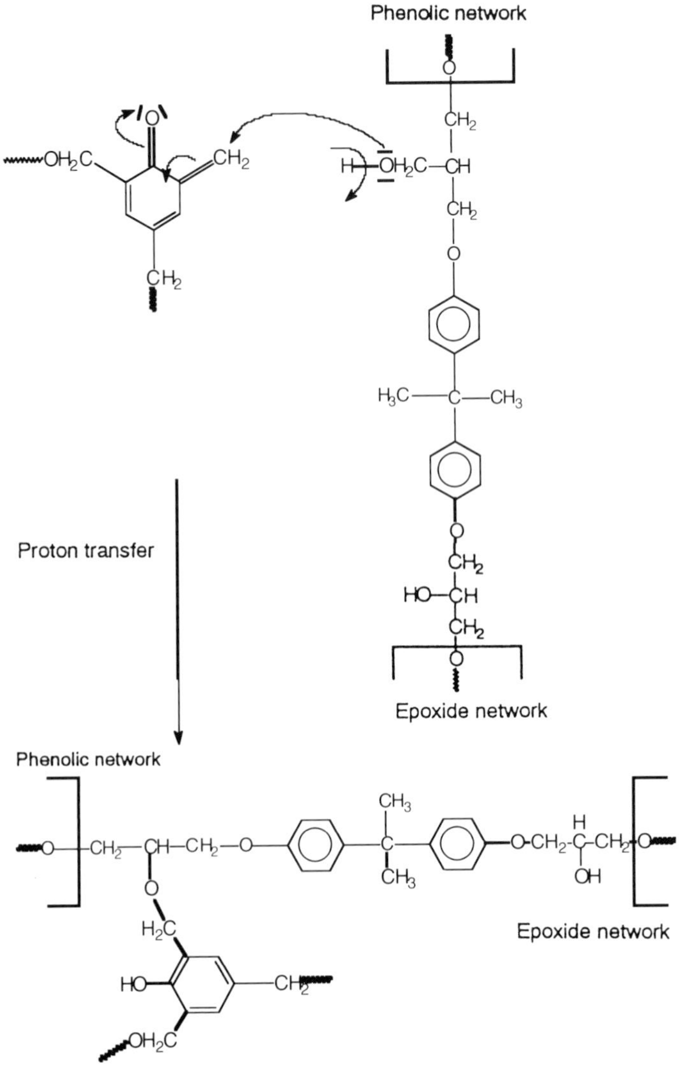

Figure 36 Reaction of the Quinone Methide structure with the secondary Hydroxyl Group via an addition reaction

Further plausible mechanisms which play a less important role are the ring-opening reaction of the residual epoxide groups via an addition reaction of a phenolic hydroxyl group, hydroxy methyl group respectively or secondary hydroxyl group, which is shown in Figure 31.

(b) *Structure property relationships for Epoxy Phenolic systems*

Epoxy phenolic coatings have to satisfy four main requirements, namely;

- high adhesion to the substrate

- high flexibility

- resistance to sterilisation

- chemical resistance

Chemical resistance, and resistance to sterilisation can be achieved by forming a highly branched, dense 3 dimensional network. The resulting network, however, will be in conflict with the second requirement – high flexibility. As a generalisation, the greater the flexibility, the more loosely the network should be bound. Flexibility and chemical resistance are a function of the degree of crosslinking, as well as the polarity of the resulting network. It is evident from Section 5 that the etherification reaction of the pendant hydroxyl groups of the epoxy and the hydroxy methyl groups of the phenolic resin lead to branching. By reducing this kind of reaction, flexibility will be increased. Different resoles, etherified with alcohols (for example Phenodur PR 612 and Phenodur PR 401®, all available from Vianova resins GmbH) are highly flexible due to their reduced reactivity towards '7' and '4' type epoxy resins

Independently from the different reaction mechanisms which have to take place to reach desired network structures, the resulting network should have enough structural elements to make contact with the surface of the substrate to enable adequate interfacial adhesion between polymer and metal.

For coatings with good initial adhesion to steel, loss of adhesion has been attributed to disruption of the coating metal interface by chemical reagents, for example humidity[40]. Humidity induced adhesion loss could also involve chemical processes in the interfacial region, but displacement by water and/or weakening of the polymer film (perhaps in the interfacial region to form a weak boundary layer) seem more probable mechanisms. For displacement to occur, the interaction between the coating and the substrate must be weaker than the attraction between water and the substrate. In this case, adhesion loss would result in the generation of interfacial metal substrate surfaces having little or no coating residue.

Adhesion of polymer to metal substrates has been claimed to result from the interaction of the functional groups of the polymer with the oxide layer present on the surface of most metals[38]. It has been proposed that this interaction involves acid base interactions, such as hydrogen bonding between functional groups of the polymer and the metal oxide. Other interactions, such as dipole-dipole interactions, have been claimed to play only a small role in maintaining adhesion[38,39]. The resistance to adhesion loss in epoxy phenolic systems is, in comparison to other coating systems (for example, polyurethane or acrylic systems), very high. This is due to the high content of free hydroxy groups which are not consumed during the crosslinking process. These hydroxy groups are available for interaction with the metal substrate and will therefore improve adhesion. The adhesion of epoxy phenolic based coatings may also be enhanced by the interaction of the electrons of the aromatic rings in epoxy and phenolic resins with the metal sites of the oxide surface. Therefore, the aromatic rings of the epoxy phenolic system act as a base and the metal species of the oxide acts as an acid. The resulting acid base interaction could account for the improved adhesion observed with aromatic systems. In general, to get excellent adhesion of the epoxy phenolic film to the substrate, the coating must also have a lower surface tension than the substrate[42].

7. Waterborne Phenolic Systems. Dispersions

Increased interest in pollution abatement and in providing higher molecular weight phenolic resins as aqueous systems, led to the development of new processes in the manufacture of phenolic dispersions. These systems are resin-in-water and resin-in-non-solvent mixtures, in which the resin exists as discrete particles. Particle size ranges from 0.1 to 2 μm for stable dispersions. The process involves the preparation of phenolic resins as stable, discrete, spherical particles in water as the continuous phase. The principal factor is the incorporation of a protective colloid which allows in situ dispersion.

The process involves a primary reaction cycle under basic conditions at 90–105°C and a secondary reaction cycle at 80–90°C under neutral conditions in the presence of a protective colloid which stabilises the phenolic particles and prevents agglomeration. During the primary cycle, condensation of formaldehyde and phenolic monomer is the main reaction. In the secondary cycle, inversion from a water-in-phenolic dispersion to a phenolic-in-water dispersion takes place. In the primary cycle amines are used as catalysts[30]. Some examples of protective colloids are poly(vinyl alcohol), cellulosic derivatives, natural gums, or combinations of these with an emulsifying agent [31,32].

Besides the in situ process, there is a post dispersion process, in which the phenolic resin has already been prepared. Solid phenolic resin is added to a mixture of water, co-solvent and dispersant, with very high shear mixing and almost always some heating. The co-solvent (frequently an alcohol or glycol ether), and heat, soften the resin, allowing small particles to form. On cooling, the resin forms particles, stabilised by dispersant and perhaps thickener, hardener and anti-settling agents. Both resole and novolac resins have been made by this process[33].

Irrespective of the dispersion process used, the protective colloid always has the same function, which is assisting particle formation and stabilisation of the particles, once formed, against coalescence. This kind of phenolic dispersion, known as a non-ionic emulsion, has three main disadvantages, namely:

- The particle size is, in comparison with ionic stabilised systems, very large. The larger the particle size, the poorer the flow and wetting properties. Therefore a dispersion with a small particle size will have better porosity (a measure of the integrity of a film, by an electrical technique, used in the can coating industry) than a dispersion having a larger particle size.

- Ionic stabilised systems can be pigmented more easily than non-ionic stabilised systems.

- Storage stability, independently of temperature ($+50°C$, $-5°C$), for ionic stabilised systems is much higher than that of non-ionic stabilised systems.

A new class of waterborne phenolic epoxide resin precondensates, which can be used as a sole binder in interior can coatings, has been developed recently[34]. This consists of two different resins brought together via a pre-condensation process. The novelty is that the co-cure resin (the phenolic resin) has already been co-reacted with the epoxy resin and despite this precondensation a highly reactive and flexible product, compared to conventional resoles, has been developed. This type of pre-condensation is a polyaddition reaction of a polyalkylidene polyphenol and a '4' or '7' type epoxy. The polyaddition reaction is carried out in the molten state at 130°C under basic conditions (for example, tertiary amines or quaternary amonium salts). The progress of the polyaddition reaction can be monitored through the viscosity, melting point and its concentration of epoxide groups, which approaches zero as the reaction porgresses, as controls. Following pre-condensation, the polyaddition product is diluted in suitable solvents, such as ketones or alcohols. Examples of suggested ketones are acetone and methyl isobutyl ketone and alcohols, propanol or butanol. In a second process, an anionic self-crosslinking phenolic epoxide pre-condensate can be obtained by the introduction of carboxyl groups and, optionally, methylol groups. To achieve this reaction product from the first process step is reacted in aqueous solution with one or more halocarboxylic acids and, optionally, before, during or after the reaction with formaldehyde. The reaction with formaldehyde or halocarboxylic acids is best carried out in basic conditions (for example, sodium or potassium hydroxides). After neutralisation with inorganic acids, for example sulphuric acid, the salts produced in the second reaction step are washed out with water after the reaction. A salt free, phenolic epoxide product can be obtained as an aqueous dispersion by converting it into a salt form. This is carried out using nitrogen bases, such as ammonia or amines, preferably triethanolamine. After neutralisation, any solvents used in the preparation are removed by vacuum distillation using toluene or xylene as entrainer.

The polyalkylidene polyphenol used for the polyaddition reaction with the epoxide has a degree of condensation of 3 to 6 and number average molecular weight $\bar{M}_n$ from 300 to 600. The average structure is shown in Figure 37.

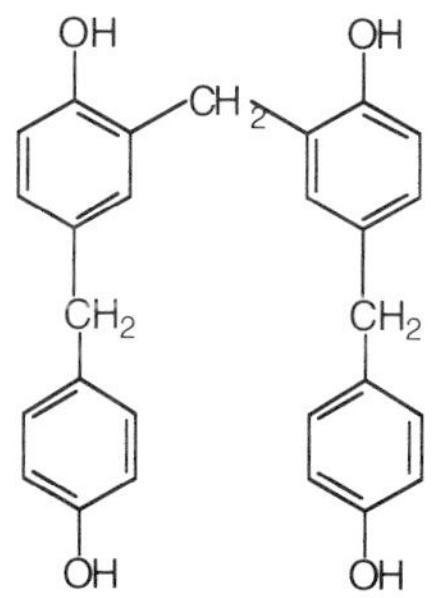

Figure 37   Idealised structure of a Polyalkylidene Polyphenol for precondensation with epoxy resin

The epoxide compound has, on average, two epoxide groups per molecule and an average epoxide equivalent weight of 800 to 4000. The number average molecular weight $\overline{M}_n$ is generally between 1600 and 8000. In the first step of the reaction addition of the phenolic hydroxyl group with the epoxide group will occur, as depicted in Figure 38.

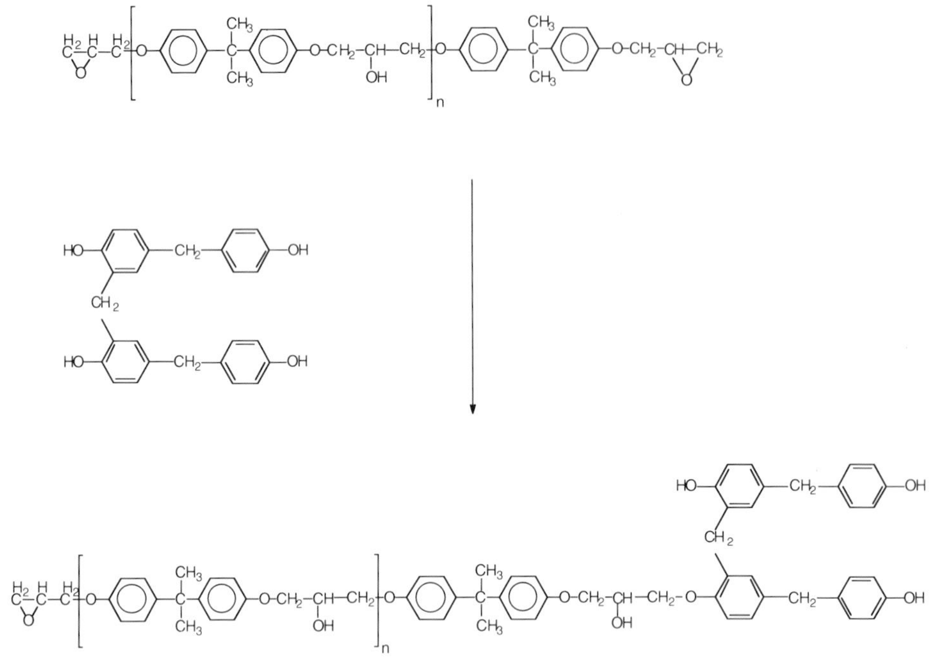

Figure 38 Polyaddition reaction of Epoxide with polyalkylidene polyphenol

The introduction of formaldehyde as well as carboxyl groups can be illustrated as Figure 39.

The idealised structure of the neutralisation of the carboxy groups, through the addition of an amine, can be seen from Figure 40.

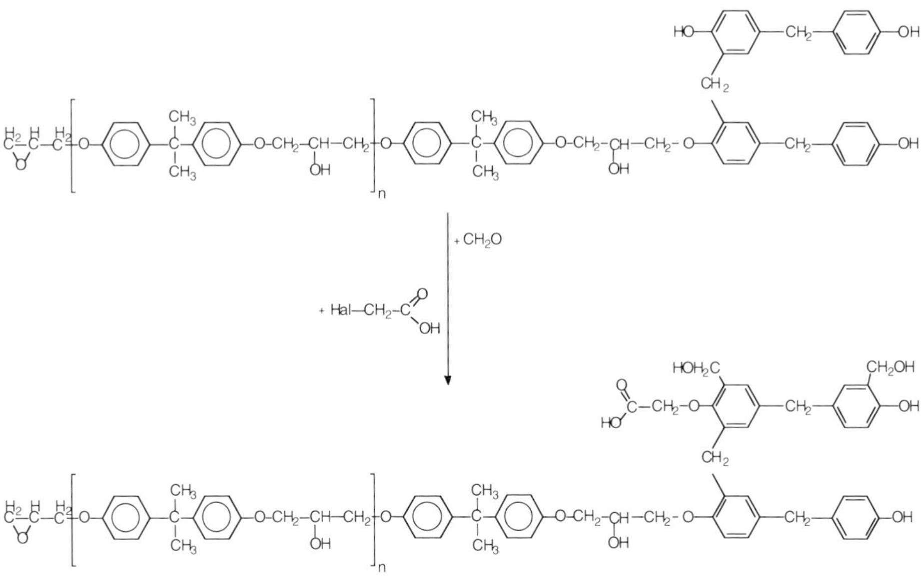

Figure 39 Introduction of Formaldehyde and Halocarboxylic Acids in the pre-condensation product of Polyalkylidene Polyphenol and Epoxide

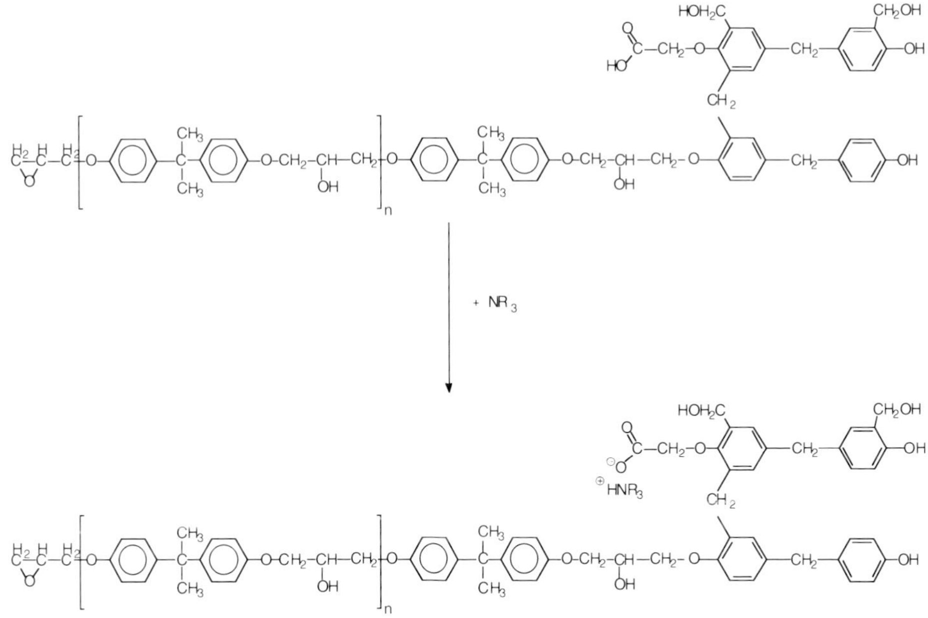

Figure 40 The idealised structure of a neutralised Polyalkylidene Polyphenol Epoxide adduct, through the addition of an Amine

(a) *Phenol*

Mononulclear phenol is a highly toxic, protein-degenerating compound. The Threshold Limit Value (TLV)-Time Weighted Average (TWA) is 5 ppm. Phenol is absorbed by the skin and causes severe blistering. Contact with large quantities can cause paralysis of the central nervous system and, ultimately, death. The destructive effect of alkyl phenols on the skin, compared to phenol, is reduced. An excellent review summarising enviromental and health risk of phenol is recommended[35] to all readers.

(b) *Formaldehyde*

Aqueous formaldehyde can cause severe irritation in contact with skin, eyes or the nose and throat. The National Institute for Occupational Safety and Health (NIOSH), USA, permits formaldehyde exposure in the air up to 3 ppm with a ceiling concentration of 5 ppm. In Western Europe exposure levels in the work facility vary from country to country with a TLV of 1 ppm being common in many Europen countries (for example, Western Germany and Switzerland).

(c) *Regulation*

The sixth Amendment of the EU rules that all phenolic resins containing more than 5% mononuclear phenol have to be marked "poisonous". Phenolic resins containing 1–5% free phenol are considered detrimental to health and are to be designated with a "St. Andrew's cross". Phenolic resins with a concentration of free phenol less than 0.2% are considered non-toxic. The label of the package or container containing phenolic resins should provide the user with the following information;

- Type of toxic components
- Warnings of special dangers
- Safety precautions

(d) *Recent trends*

The result of many toxicological experiments involving phenolic resins (resoles) with a content of mononuclear phenol greater than 20%, is not in agreement with the classification of phenolic resins. There is a major difference between the toxicological behaviour of phenolic resins and phenol itself[36]. Although these differences exist, the reduction of mononuclear phenol and formaldehyde are very important, due to their possible emissions at the work place (MAK-value: Maximum Concentration at Place of Work) and environment. Many efforts are being made to develop new phenolic resins, especially resoles, as co-cure resins which are completely free of mononuclear phenol and formaldehyde respectively, for example waterborne phenolic resins.

8. Toxicology

The toxicology of any raw materials used together with products manufactured from them must never be overlooked. Many of the raw materials used in the manufacture of phenolplasts are hazardous. The resulting phenolic resins are manufactured to minimise any risk involved in their use. By the time, needless to say, coatings have been converted into a cured film, any residual raw materials remaining in the film are minimal. In the case of food contact the small amounts of these raw materials which could remain are heavily regulated.

References

(1)   Von Bayer, A.: Chem. Ber., **5**, 25,280,1094 (1872)

(2)   Baekeland, L.H.: US-PS 942 699 (1907); DE-PS 233 803

(3)   Hultzsch, K. (1950) *Chemie der Phenolharze*, Berlin, Göttingen, Heidelberg, Springer.

(4)   Martin, R.W. (1956) *The Chemistry of Phenolic Resins*, New York, J. Wiley.

(5)   Megson, N.J.L. (1958) *Phenolic Resin Chemistry*, London, Butterworths.

(6)   Pujado, P.R., Salazar, J.R., Berger, C.V.: Hydrocarbon Proc., 91 (1976)
      Seubold, F.H., Vaugham, W.E.: J. Amer. Chem. Soc. **75**, 3790 (1953)

(7)   Fiege, H. et al: "Phenol." In: *Ullmanns Encyclopädie der Technischen Chemie*, Vol. **18,** 191 (1979), 4. Ed. Weinheim: Verlag Chemie
      Clouts, K.E., McKetta, R.A.: "Cresols and Cresylics Acids." In: McKetta J.J. Ed: *Encyclopedia of Chemical Processing and Design*, Vol. **13**, 212 (1981), New York: Marcel Dekker Inc.

(8)   Ito, K.: Hydrocarbon Proc., 89 (1973)
      Reed, H.W.B.: "Alkylphenols." In: Kirk-Othmer, *Encyclopedia of Chemical Technology*, Vol. 2, p. 72, New York: J. Wiley & Sons (1978)

(9)   Diehm, H., Hilt, A.: "Formaldehyde." In: *Ullmanns Encyclopädie der technisachen Chem.*, **11** (1976), 4. Ed. Weinheim: Verlag Chemie

(10) Maux, R.: "Air Best for Formaldehyde and Maleic." *Hydrocarbon Proc.*, 90 (1976)
      Diehm, H., Hilt, A.: In: *Ullmanns Encyclopädie der Techn. Chem.*, Vol. 11, 4. Ed., Weinheim: Verlag Chemie 1976.
      Diehm, H.: Chemical Enging. 27, 83 (1978)

(11) Walker, J.F.: Formaldehyde, ACS Monograph No. 159, 3 Ed. New York: Reinhold Publ. Co. 1964
      Diehm, H., Hilt, A.: In: *Ullmanns Encyclopädie der Techn. Chemie.*, Vol. 11, 4. Ed., Weinheim: Verlag Chemie 1976.
      Diehm, H.: Chemical Enging. 27, 83 (1978)

(12) Zabicky, J.: *The Chemistry of the Carbonyl Group*. London: Interscience 1970

(13) Knop, A., Pilato, L.A.: *Phenolic Resins* p. 27. Springer Verlag: Berlin Heidelberg New York Tokyo (1985)

(14) Yeddanapalli, L.M., Francis, D.J.: Makromol. Chem. **55**, 74 (1962)

(15) Sprung, M.M.: J. Am. Chem. Soc. **63**, 334 (1941)

(16) Haller, D., Schmidt, K.H.: Diplomarbeit T.H. Essen, 1983

(17) Whitehouse, A.A.K., Pritchett, E.G.K., Barnett,G.: *Phenolic Resins*, London Iliffe Books Ltd: Plastics Institute p. 27 (1967)

(18) Hesse, W., Hofmann, K., Leicht, E.: US Patent 005089589 A

(19) Fry, J.S., Merriam, C.N., Boyd, W.H.: Applied. Poly. Science, ACS 1985

(20) Cangelosi, F., Shaw, M.T.: Polym. Plast. Technol. Eng. **21**, 13 (1983)

(21) Pennaccia, J.R., Pearce, E.M., Kwei, T.K., Bulkin, B.J., Chen, J.P.: Macromolecules **19**, 973 (1986)

(22) Olabisi,O.: Macromolecules **8**, 235 (1975)

(23) Whitehouse, A.A.K., Pritchett, E.G.K., Barnett,G.: *Phenolic Resins*, London Iliffe Books Ltd: Plastics Institute P. 67 (1967)

(24) Knop, A., Pilato, L.A.: *Phenolic Resins* p. 148. Springer Verlag: Berlin Heidelberg New York Tokyo (1985)

(25) Knop, A., Pilato, L.A.: *Phenolic Resins* p. 252. Springer Verlag: Berlin Heidelberg New York Tokyo (1985)

(26) Bourlier, K.: Journal of Coatings Technology Vol. 68 No. 853 p. 59 (1996)

(27) Yamamoto, K.: Chap. 11, ACS 1994

(28) Shechter, L., Wynstra, J.: Ind. Eng. Chem. **48**(1), 86 (1956)

(29) Smith, M.E., Ishida, H.: Macromolecules 27. 2701 (1994)

(30) Dijkstra, R.: U.S. Pat. 3,843,575 (1974)

(31) Hofel, W.B., Kiessling, H.J.: US Pat. 3,878,136 (1975)

(32) Ingram, W.H.: US Pat. 3,666,694

(33) Fry, J.: US Pat. 4,124,554 (1978)

(34) Gräff, H., Gräff, R., Has, A., Hesse, W.: US Pat. 5177161 (1993)

(35) Babich, H., Davies, D.L.: Regul. Toxicol. Pharmacol. **1**(1), 90 (1981)

(36) "Bakelite Phenolharz – Toxicologische Eigenschaften": Firmenschrift der Bakelite GmbH, Iserlohn 10/89

(37) Fedtke, M., Domaratius, E.: Plaste und Kautschuk **12**, 434 (1985)

(38) Fowkes, F.M., Mostafa, M.A.: Ind. Eng. Chem. Prod. Res. Dev. **17**, 2 (1978)
Fowkes, F.M.: In: "Adhesion and Adsorption of Polymers", Lee, L.H., Ed.: Plenum: New York, 42 (1980)

(39) Fowkes, F.M., Sun, C.Y., Joslin, S.T.: In "Corrosion Control by Organic Coatings", Leidheiser, H.: Ed.: National Association   of Corrosion Engineers: Houston, Texas, 1 (1981)

(40) DeVries, J.E., Holubka, J.W., Dickie, R.A.: Ind. Eng. Chem. Prod. Res. Dev. **22**, 256 (1983)

(41) Houben-Weyl, 4th. Ed. XIV12, 272 (1985)

(42) Payne, K.L., Puglisi, J.S.: J. Coat. Technol., **59**, 117 (1987)

CHAPTER III

By P. Oberressl

APPLICATION OF PHENOPLASTS IN COATING SYSTEMS

Chapter III

1. Introduction

Although the amount of phenolic resin used in the coating industry is, when compared with the amount used in other Industries (wood binding, foundry industry), relatively small, it is estimated that in Europe, alone, approximately 20,000 metric tons of solid phenolic resins are used in coating systems. The bulk of this tonnage ($> 50\%$) is used for varnishes for electrical components, such as motor windings, and wire.

Most of the remainder is used in two different coating applications,

- Corrosion protection coatings
- Coatings for metal packaging

In both applications, phenolic resins (resoles in most cases) are combined with other binders such as epoxides, poly vinyl butyral, polyester, acrylic or PVC resins. The phenolic resin part of the resin film forming combination is that which gives a high degree of crosslinking, film hardness and chemical resistance. In most cases (there are however notable exceptions), phenolics are used as co-curing resins. This means, that not only are they fundamental to the curing (film forming) process, but also have film forming properties in their own right.

2. Corrosion Protection

The need to protect steel and other non-ferrous metals from corrosion, motivated intensive research for suitable binder systems for corrosion protective coatings. There are, in the meantime innumerable ways and means of protecting metal from corrosion. The applicant can find tailor made systems for practically all possible circumstances. This section of the book deals only with the small and special section of phenolic resins as co-curing resins (or binders)for air drying and stoving corrosion protection systems.

Special resoles have played a prominent role in corrosion protective primer coatings for a very long time and they are still indispensable in many cases. One of the most important corrosion protection systems is the wash or shop-primer. The main features of this type of primer are universality and ease of use (for all possible substrates, and for industries such as ship building and industrial corrosion protection, rapid drying and excellent corrosion protection.

The most important needs for such wash – or shop – primers are:

- rapid drying, overcoatability after a few minutes
- thin films (5-15 grams/m^2 or approximately 3-10 µm dry film thickness)

- excellent corrosion protection properties, even if not overcoated for prolonged periods of time

- excellent adhesion to all ferrous and non-ferrous metals

- weldability (producing little or no noxious effluents during welding processes)

- good storage stability (particularly for one pack primers), and no pigment settling

Coating systems are often referred to as one or two pack. In a two pack system, it is necessary to mix two components together before the application of the coating. This may be a hardener and a second resin system, stored in two containers until ready for use. It could involve the addition of a small component such as a catalyst. The storage stability of the coating is the reason for mixing prior to use. The length of time during which the coating remains fluid enough to apply (or perform, once applied)is called pot life.

On the other hand, a one pack or one pot system is one in which all components are mixed at the coating manufacurers and are in essence used as supplied, with perhaps the exception of additions of solvent. Obviously, having all of the reactants mixed together means that the coating formulator has to verify that the coating will still be usable after a certain period of time, the shelf life, which is normally a minimum of 6 months and often 1 year.

Phenolic resins are seldom used as the sole binder in corrosion protection systems. In most cases, phenolic resins are combined with one or more other resins to achieve the required properties. The phenolic resins discussed here are of the reactive resole type which show a high degree of reactivity if catalysed with acids. The resoles used in this application are of a relatively low molecular weight ($\overline{M}_w$ approximately 2000) and have good compatibility with various non-phenolic binders. A suitable binder combination would be like the one given in Formulation 1.

FORMULATION 1 : WASH/SHOP PRIMER BINDER COMBINATION

MATERIALS	wt.% solid resin
Phenolic resin (e.g. Phenodur PR 263)	8%
Low molecular weight epoxide resin	9%
Polyvinyl butyral (e.g. Mowital B 30H)	9%
Phosphoric acid	0.2-0.7 (100%)

Solvents and additives and, in the case of pigmented systems, anti corrosive pigments are added.

(The amount of phosphoric acid basically determines, whether it is a one-pot or a two pot primer)

Different binders have different functions. In such a mixture, the phenolic resin is responsible for chemical resistance and the integrity of the film. The low molecular weight epoxide resin is responsible for excellent adhesion and the flexibility of the film. The polyvinyl butyral is responsible for quick drying, good film formation and, having also a certain potential degree of crosslinking with acids, good corrosion protection.

Phosphoric acid is itself a very important part of the binder system. Indeed the whole principle of wash-priming would not be possible without it. On the other hand, if much more than the amount mentioned above is added, one would face adhesion problems, discolouring and a deterioration in corrosion protection.

(i) **Substrates**

Virtually all sorts of metal substrate used in construction may be primed with a binder system similar to that given in Formulation 1 which is based on a resole phenolic resin. The most important substrates are listed in Table 1:

TABLE 1 : POSSIBLE SUBSTRATES

• Steel
• Phosphated steel (iron phosphate)
• Bonderized steel (zinc, or other phosphates)
• Zinc plated steel (electrolytically or sendzimir)
• Aluminium
• Other non ferrous metals or alloys (brass, bronze etc.)

In practice, cold rolled plain steel is very often used un-cleaned The surface contains grease, lubricants and superfluous corrosion. Especially in big construction industries, such as the ship building industry, there is no practical way to clean the steel surface thoroughly. So, one of the most important requirements of a wash- or shop primer is to adhere perfectly on un cleaned, even already slightly corroded steel surfaces.

Phosphated steel is much less critical, although the treated surface is often rather acidic (from residue salts stemming from the pre treatment process). Zinc plated steel is again rather difficult to coat, since the actual adhesion of a given system depends upon the age of the zinc layer. As a basic rule, the 'younger' the zinc plate, the more difficult it is to achieve perfect adhesion. Furthermore, if the zinc coating of the zinc plated steel is defective (as is done deliberately during a regular hot salt spray test or a humidity cabinet test), a local element is formed immediately and current induced corrosion takes place, stressing the primer coat even more.

Aluminium is the least critical surface to be primed, and the most difficult challenge is adhesion to the numerous aluminium alloys in use. Other alloys, such as brass or bronze, are seldom used in construction, examples are window wiper constructions for the automotive industry, where at least 3 different metals and even a plastic (often polyamide) have to be primed with one system. Primers based on phenolic resins discussed in this chapter are a perfect way to acheive this.

(ii)   **Application methods**

Wash – or shop – primers are applied in many ways, depending upon whether they are used in industrial finishes , DIY or by professional painters. Methods of application are given in Table 2:

TABLE 2 : APPLICATION METHODS

• Dipping
• Spraying (air assisted, airless, airmix)
• Brushing
• Flooding

(iii)   **Introduction**
Fast to fast drying primers

Such systems have been in use for many years. Their usefulness stems from their quick drying properties and from the fact that they can be overcoated as soon as they are dry. In certain industries, such as ship building, pre-fabricated parts are used to a great extent to shorten building or repair (wharf) times. Previously pre-fabricated steel parts have to be protected against corrosion, at first temporarily and then, after they have been built into the structure, permanently. The priming system used has to serve both purposes and resole based wash or shop primers fit these demands exactly. They dry rapidly, and have within themselves (without being overcoated) a sufficient degree of corrosion protection (at least for a few months at thicknesses of approximately 5-10µm) If overcoated, they have a significant protection time.

Some properties of Formulation 2 (opposite) are:-

Solids (by weight)   17.58%

Phenolic resin : PVB   approx. 1:1

Viscosity DIN 53211/4mm/23°C   approx. 20″

Touch free drying time (10µm dry film/23°C)   approx. 10 minutes

**FORMULATION 2 : CLEAR ETCH PRIMER BASED ON POLYVINYL BUTYRAL AND
RESOLE PHENOLIC RESINS**

MATERIALS	Wt.%	solid %
Phenodur PR 263 / 70%	12.20	8.54
Mowital B 30 H/20% solution*	45.20	9.04
Thinner **	39.30	
Phosphoric acid 85%/1:3 in butanol	3.30	
	100.00	**17.58**
* Mowital B 30 H solution:		
Mowital B 30 H	20.00	20.00
Butanol	20.00	
Methoxy propanol	20.00	
Xylene	40.00	
	100.00	**20.00**
** Thinner		
Butanol	30.00	
Methoxy propanol	30.00	
Xylene	40.00	
	100.00	

Formulation 2 is typical of a fast drying, clear etch primer. First, the polyvinyl butyral solution is prepared by dissolving the PVB powder in the solvent mixture. Care should be taken so that no undissolved, jelly like particles are visible. The best way to avoid this is to premix the powder in xylene and to keep this slurry under constant mixing whilst adding the other diluents.

The phenolic resin, in this case, a very reactive acid curable resole (Phenodur PR 263 70%, is added afterwards. Phosphoric acid is always added last. The storage stability of such a primer is approximately 6 months. Although the corrosion resistance is somewhat lower than that of a pigmented version (see below), reasonably good results in salt spray and high-humidity cabinet tests are obtained. A small amount of water is needed to promote the catalytic action of the phosphoric acid. There is however no need to add water, since phosphoric acid of 85% strength is used.

Formulae 3 and 4 are both for the same purpose, their sole difference is in the use of a chromate free anti-corrosion pigment in the case of formula 3 in which a zinc phosphate pigment, instead of the zinc chromate pigment in formula 4 is used. Phenodur PR 263 is the reactive resole used in both formulations. Pigments and fillers are dispersed in the phenolic and epoxide resins plus an appropriate amount of solvent to obtain the required grinding viscosity. This again depends on the grinding equipment. The fumed silica of which a pre-paste is made with the help of an anionic pigment wetting agent (for example, Additol XL

FORMULATIONS 3 AND 4
PIGMENTED WASH-PRIMER, CHROMATE CONTAINING AND CHROMATE-FREE

MATERIALS	wt. %	wt. %
Mowital B 30 H/20% IPA/X/Methoxy propanol	36.00	36.00
Phenodur PR 263/70%	9.00	10.00
Beckopox EP 301/50%MP	9.00	10.00
Talc powder IT ex. (Norwegian Talc)	9.00	9.00
Bayferrox 120M (Bayer)	5.00	5.00
Heucophos ZPO (Heubach)	10.00	
Zinc chromate KSHSM (Heubach)		8.00
HDKH15 (fumed silica)/10% in Xylene		
(+5%Additol XL 270)	4.00	4.00
(Wacker)		
Thinner (X/IPA/MP 34:33:33)	15.00	15.00
10 H3PO4 85%/1:3 in butanol	3.00	3.00
	100.00	**100.00**

pot life: chromate free about 1 month, containing chromate pigment about 6 months

270), has the function of an anti-settling agent. At the same time, it acts as a 'pore filler', because of its extremely low primary particle size (< 10 nm). The use of montmorrillonites is not recommended, because these types of anti-settling agents have, in some cases, a negative influence on the corrosion protection properties.

This primer may, under 'normal' conditions, (for example, $20°C$) be overcoated after only a few minutes. At a film thickness of 10 μm dry film, this primer is tack free after only 5 to 10 minutes at room temperature (approximately $23°C$).

(iv) **Stoving systems**

Phenolic resins find usage in stoving systems. Sometimes, for industrial applications, corrosion protection primer or systems are required to cure at high temperatures. Very often, phenolic resins are combined with polyester or alkyd resins. In some cases, where flexibility is not required, phenolic resins can even be used as the sole binder. Some examples are given in Table 3.

This primer as a one coat system offers excellent corrosion protection on untreated, but cleaned, cold rolled steel. At 15 μm dry film thickness, stoved for 20 minutes at $190°C$, the coating system is resistant under hot salt spray test conditions for at least 1000 hours.

Phenodur PR 404 is a reactive resole. The pigments used should be dispersed into both the alkyd and the phenolic resin on a ball mill, to the required fineness of grind. Care should be

TABLE 3 : EXAMPLES OF STOVING SYSTEMS CONTAINING PHENOLIC RESINS

- Coatings for shock absorbers, oil filters
- Coatings for window wipers
- Chassis coatings
- Axle coatings
- Interior coatings for air conditioner
- Coil coating primer
- Coil coating backing systems

FORMULATION 5 : STOVING PRIMER OR ONE-COAT SYSTEM, RED

MATERIALS	wt. %	solid
millbase		
Short oil alkyd resin/60%xylene	30.00	18.00
Phenodur PR 404/55%	42.50	23.40
Iron oxide red micronised	9.00	9.00
Aerosil R 976	0.30	0.30
let down		
n-Butanol	1.50	
Solvesso 100	7.90	
Methoxypropanol	8.00	
Additol XL 122 (flow agent)	0.40	
Additol XL 480 (flow agent)	0.40	
TOTAL	**100.00**	**50.70**

Viscosity: DIN 53211/4mm/23°C :approx. 160"
Diluent : Sol.100/But./MP : ca. 1:1:1
Typical stoving cycle (depending on the metal thickness) 15 – 30 min. / 180 – 200°C

taken to ensure that the temperature during the grinding process does not exceed 50°C for any length of time.

Aerosil R972 is a fumed silica used as anti settling agent in this formula. Two flow agents, one silicone free (Additol XL 480) and the other with silicone (Additol XL 122), are used to improve flow and surface wetting.

FORMULATION 6 : WATERBORNE COIL COATING SYSTEM, BACKING COAT, BLACK

MATERIALS	wt. %	solid
Phenodur VPW 1942 / 56WA	63.00	34.65
Hexylglycol	3.00	
Printex G carbon black (Degussa)	5.00	5.00
Butyl glycol	5.00	
Maprenal MF 900 (HMMM)	2.00	2.00
Di methyl ethanol amine	0.20	
Water, deionised	21.80	
TOTAL	**100.00**	**41.65**

Some properties of Formulation 6	
Volatile organic content	approx. 10% on weight
Typical solid content at application:	40 %
Typical viscosity DIN/ISO 53211/23°C	60-80s(at application)
Typical stoving schedule	60 seconds at 240°C (PMT)
Gloss, 60°	approx. 80%
Cross cut on cold rolled steel	GT 0
Salt Spray Resistance HSS (40°C)	$> 120^h$

This formula contains Phenodur VPW 1942, as a waterborne phenolic epoxide pre-condensate, which is anionic stabilised. The product has a very small particle size being less than 100 nm. The advantages over non ionic stabilised systems are:

- much better pigment wetting properties

- better flow and surface wetting

- no emulsifier, therefore less affected by water/chemicals

A hexamethoxy methyl melamine resin is used as flow agent to improve film formation and reduce 'bubbling' tendencies. This formulation contains approximately 7% on weight of carbon black. This rather high amount is needed, to get the required hiding power at low film weights of 5–10 grams/m^2. The carbon black is dispersed directly into the phenolic epoxide pre-condensate on a sand mill with at least 35m/s^{-1} circumference speed. Care must be taken to ensure that the temperature of the batch does not exceed 45°C. Some of the amine is going to evaporate under these circumstances, therefore the pH value should be re-adjusted after the grinding process to a level of 7.5–8.5.

This system has excellent adhesion onto cold rolled steel and aluminium and the coating shows gloss values of >80% measured at an angle of 60°. This can be achieved without the need of wetting agents, which might reduce corrosion protection and chemical resistance.

There is very little foam development during production or application, so there is no real need for a defoamer. In case of excessive foam formation, the addition of a small amount of white spirit as a defoamer is recommended. The hexyl glycol and the butyl glycol are both used here as film forming, flow and wetting promoting solvents.

(v) Test methods

(a) *Corrosion*

In evaluation work, paint technologists have the task of formulating workable systems. It is not possible for coating properties to be evaluated in 'normal' environments and under 'normal' conditions. For this reason, a number of accelerated tests have been developed to shorten evaluation time. This is of special importance for corrosion protection systems, because no one can wait for 5 years to tell whether a coating is suitable for a particular application or not.

The most important test for corrosion protection is still the Hot Salt Spray Test (HSST)[1]. It is carried out in a closed cabinet, in which the coated surface of the test specimen is attacked by a hot salt fog of a given potassium chloride concentration. This is an extreme but necessary exaggeration of reality. Corrosion protection systems are nowadays much better and longer lasting than they were. To increase the impact of the salt solution, the coating is partially destroyed or cut by a sharp knife. This is done to estimate the ability of the coating to prevent the aggressive salt solution from undercutting the film.

In many cases, the primer itself is tested as well as the complete coating system. It is strongly recommended that a 'standard' system of known corrosion resistance is tested alongside the developmental material being evaluated.

Another important test is the Humidity Cabinet Test[2] where condensed water and water vapour due to high humidity attack the coated panels. The test is normally done at 40°C in a closed chamber. Care must be taken to ensure that the whole surface of the test specimen is covered by the condensed water, so that the panels must not be placed too near the walls, opening or rear of the test apparatus. The main reasons for this tests are given in Table 4.

TABLE 4 : SOME REASONS FOR TESTING CORROSION RESISTANCE

• can the coating withstand osmosis?
• can the coating adhere under wet conditions?
• does the coating contain water soluble matter?
• does the coating take up water or is it porous?
• to what degree can the coating system itself prevent corrosion?

For the salt spray and the humidity cabinet tests, much care must to be taken to ensure that the conditions are similar for ;

- film weight

- drying conditions

- ageing (of cured film before testing)

- substrate

Basically, four different phenomena may be observed during this test, namely ;

- corrosion, taking place at the cut itself

- corrosion taking place between the coating and the metal near the cut (that is, salt solution undercutting the coating)

- corrosion taking place between the coating and the metal not connected with the cut

- bubbles of different size and numbers dispersed more or less evenly over the whole surface

An example of such a test with its result is given in Figure 1, where there is a direct dependence between duration and the amount of corrosion on the one hand and, on the other, the chromate containing primer is superior to the primer containing zinc phosphate as an anti corrosive pigment. The two primers corresponded to those used in Formulations 3 and 4.

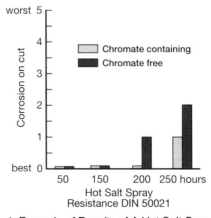

Figure 1 Example of Results of A Hot Salt Spray Test

The substrate has been cold rolled and degreased but otherwise it is untreated steel. The primers have been spray applied at a film thickness of approximately 15 μm dry film and the primers were allowed to dry for one week at room temperature (approximately 23°C).

(b)   *Other, Additional Test Methods*
The following tests are of additional importance to estimate the ability of a corrosion protection system:

Crosscut	to estimate adhesion before and after short cycle tests;
Pendulum hardness	before and after short cycle tests and at intervals during the normal drying time;
Overcoatability	with different topcoat systems (for example, alkyd, poly-urethane, acrylic resin based topcoats)
Drying time	to estimate the shortest possible time, after which the primer can be overcoated;

3.   The Can Coating Industry

This section of the chapter in principle deals with the coating of the interior of metal containers used in the light metal packaging industry. Preventing foodstuff from decay is one of the oldest tasks of mankind. Cooling, deep freezing and freeze drying are more modern methods, whilst canning, salting and simple drying are the older ones. Everything is done once for the first time and, in many cases, the inventor(s) remain unknown, their image blurred by time. The inventor of canning however is known[3]. He was Nicolas Francois Appert, a Frenchmen, born in 1749. He made his great invention approximately in 1790 by discovering that foodstuff, if put into tightly closed containers and heated up with the help of steam, stays edible for a very long period of time. M. Appert became very famous in his time, his method of preserving food (the first containers used were from glass) made his fortune.

The canning industry (first in France, then world-wide) became a very important industry, the strategic importance of which was immediately recognised by the military and politicians world-wide. The importance of preserving food by canning cannot be over estimated. Millions of people, in the future nearer billions of people will, depend on this invention and in a great number of 'underdeveloped countries', where people cannot afford refrigerators (not to mention deep freezers) they can only minimise the usually great loss of harvest, by canning.

Canning in metal containers derived from M. Appert's activities in Great Britain around 1800. Metal containers have, from the very beginning, offered great advantages over glass containers. They are lighter and practically unbreakable. However, a coating had to be developed for metal containers since contact with (mostly) acidic food, destroyed the metal and, of course, had a disastrous effect on the taste of the food. The first coatings 'English

Can-Coating's were based on resins such as sandarac, mastics, venetian turpentine and alcohol, and their resistance was rather limited. Food like fish, meat and many vegetables could not be packed in these containers.

In the year 1910, Ludwig Behrend, a chemist working for the Company, Dr. Kurt Albert in Wiesbaden, developed the first useable resins based on phenol and formaldehyde[4]. With this group of resins, constantly improved and re-designed, a suitable base was found for the successful coating of the interior of metal food containers. Up to now, phenolic resins (as curing resin) and epoxide resins (as main binder and 'plasticiser') are the most important group of binders in this field of application for food cans and closures.

Coatings based on phenolic resins are (with some notable, modern day exceptions) brownish golden in colour. The phenolic resin simply yellows under relatively high crosslinking temperatures. They became known as "gold lacquers" and this term is still used today throughout the industry.

(i)   Resins/Binders

The amount of coating materials used for coating metal packaging is enormous. It is one of the main uses inside the coating industry. Compared with most other areas of industry, there are few coating manufacturers and even fewer end users involved in this activity. Most companies are global and most of the research and development work is done for global uses and purposes. To visualise this, Figure 2 shows the estimated demand of such coatings in a number of European countries.

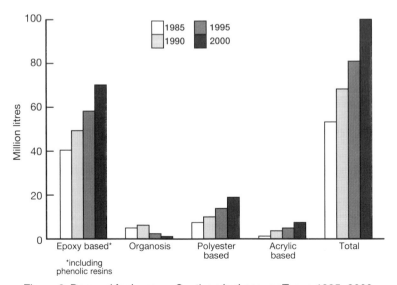

Figure 2 Demand for Lacquer Coatings by Lacquer Types 1985–2000

(a) *Epoxy Phenolic Systems*

Mixtures of high molecular weight epoxide resins, with some portion of phenolic resin as curing agent, are now seen as the 'universal' system for the internal coating of metal containers used in the food packaging and other related industries. The blending of a rigid phenolic resin with a high molecular weight epoxide resin leads to almost perfect interior can coating systems, although epoxide resins, even high molecular weight ones, are not usually considered to have high flexibility. The main reason why such coatings are flexible, is the low film thickness at which they are usually applied, which can range from:

- 1–2 grams per m^2 for sizing

- 3–6 grams per m^2 for "normal" internal coatings to

- 10–15 grams per m^2 for highly chemical resistant interior drum coatings

Phenolic resins in these mixtures give:

- chemical resistance

- sulphur staining resistance

- retort resistance

- solvent resistance

- hardness

- colour

However, phenolic resins tend to have poor taint characteristics. The epoxy resin is mainly responsible for :

- flexibility

- adhesion onto all sorts of metallic substrates

Phenolic resins could be used as the sole binder, to take advantage of their very high chemical resistance (and this is done in rare cases, where no flexibility is required). In practice the combination of a high molecular weight epoxide resin with one or more phenolic resins is used. This is since a balance of properties has to be achieved between chemical resistance, flexibility, reactivity, taste (organoleptic) properties, colour and storage stability of the lacquers, for every application and on an individual basis. This is indeed a very demanding objective for everyone active in this part of the coating industry. Phenolic resins used for this purpose are mainly:

- Resoles

- Functional Novolacs

Epoxide resins used for this purpose are mainly:

- High molecular weight types such as

- Type '7' or '9' or even phenoxy resins of molecular weights above 30 000.

The ratio for phenolic and epoxide blends depends to a great extent upon required reactivity, flexibility, resistance and adhesion properties and the substrate.

The mixing ratio may be between 10:90 and 90:10 (solid phenolic : solid epoxide resin) and for food container interior systems between 15:85 and 50:50. The mixing ratio is rather wide, mainly because different applications require different degrees of flexibility and of chemical resistance.

For example, an interior drum coating for steel drums has to have an extremely high degree of chemical resistance. All sorts of aggressive chemicals are filled into steel drums and the individual coating should withstand most of them. So, in this case, a highly chemically resistant but less flexible system is required and the resin mixture might contain up to 90% of a resole and only 10% of a flexibilising resin, such as high molecular weight epoxide resin or a poly vinyl butyral resin.

On the other hand for an interior coating of a 2 piece deep drawn fish can ('club can') a very flexible coating is required, which retains enough chemical resistance to withstand the retort process and the food for very long periods of time. In this case, a mixture of only 15–25% of a resole with 85–75 % of a high molecular weight epoxide resin might be a sensible choice.

It should be noted that whilst phenolic resins, as the sole binder, can give flexible films, if the film weight is low enough, any variation above that weight quickly gives films of poor flexibility. Film weights below this value had the potential inability to protect either can or contents. In the early days of roller coating application, film weights could not be controlled within the accuracy required, so that phenolic resins were not used as the sole binder.

(b)   *Types Of Phenolic Resins*
Compared with other synthetic resin groups used in the coating industry, the number of available phenolic resins for the can coating industry is rather limited. Far from being complete, Table 5 shows an overview of some typical grades from some different manufacturers.

For the sake of uniform comparisons, the bulk of discussion about the selection of phenolic resins will focus on the range of phenolic resins available from one supplier, namely Vianova Resins GmbH. This in no way implies that this is the only phenolic resin which will perform adequately in a given system and many phenolic resin suppliers are able to offer alternatives. However, some phenolic resins are unique. Detailed discussions with the technical representatives of a phenolic resin supplier should result in a selection of suitable phenolic resins for the evaluation in a particular application.

TABLE 5 : SOME TYPICAL RESOLES USED AS BINDER IN CAN COATING SYSTEMS

Non plasticized , non etherified resoles:	
Bakelite 100	Phenodur PR 373
Bakelite LB 737	Phenodur PR 722
Bakelite 9081LB	Phenodur PR 897
Bakelite Phenolic Resin BKS 2700	Phenodur VPR 1776
Phenodur PR 217	Phenodur VPM 1777
Uravar FB 120	Schenectady SFC 087
Uravar FB 190	Schenectady SFC 093

Non plasticized, etherified resoles:	
Epicure DX 200	Phenodur PR 285
Bakelite LG 7700LB	Phenodur PR 308
Bakelite LG 7576LB	Phenodur PR 612
Bakelite Phenolic Resin 7550	Phenodur PR 401
Bakelite Phenolic Resin BKS 7597	Phenodur VPR 1775
	Phenodur PR 515
Santolink EP 560	Uravar FB 250
Schenectady SFC 092	Uravar FB 210
Schenectady SFC 099	

Plasticized resoles (pre condensates):
Beckopox EM 524
Uravar L 35 M2-41
Varcum 500

Waterborne systems:
Bakelite Phenolic Resin BKUA 2370
Phenodur VPW 1942

(c)　*Application Related Differences Between Some Phenolic Resins (Resoles)*
Resoles used as co curing resins in can coating systems differ mainly in

- Chemical Resistance
- Colour
- Compatibility
- Flexibility
- Reactivity

Figure 3 shows differences in reactivity of some phenolic resins.

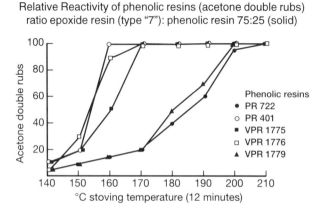

Figure 3 Difference In Reactivity Of Some Phenolic Resins

The lacquers based on the individual phenolic resins plus the high molecular weight epoxide resin were coated onto tinplate (ETP), giving a film weight of approximately 5 grams per m^2. The acetone double rub test (weight applied approximately 1kg) was chosen to determine the crosslinking density at a given stoving temperature. It was shown that there were significant differences in the intrinsic reactivity of resoles used to cure interior can coating systems. These differences can be used to formulate systems for practically all purposes.

Figure 4 shows the difference in flexibility (slow deformation) of some phenolic resins.

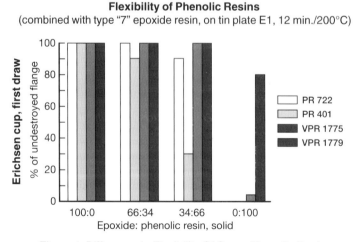

Figure 4 Difference In Flexibility Of Some Phenolic Resins

Figure 4 shows the difference in flexibility between a number of phenolic resins used to formulate interior can coating systems. Four different ratios between the high molecular weight epoxide resin and the appropriate resole have been evaluated: 100:0, 66:34, 34:66 as has the pure resole film without a high molecular weight epoxide resin. The substrate is tinplate (ETP), the film weight approximately 5 grams per m2 and the stoving cycle 12 minutes at 200°C in an air circulation laboratory oven. An Erichsen cupping apparatus, model 224/2 (slow deformation process) is used to stamp out the cups from the tinplate. The results indicate that a very reactive resole, like Phenodur PR 401, shows less flexibility than a lower reactive type like Phenodur PR 722. The higher molecular weight type resole VPR 1779 shows the highest intrinsic flexibility of all tested resoles and, of course, as soon as the ratio is altered in favour of the high molecular weight epoxide resin, the flexibility improves. As always, a compromise has to be reached between flexibility and reactivity. This sort of information is naturally very useful where the properties of cured coating have to be modified. Figure 5 shows the difference in flexibility under a fast deformation process (wedge bend test) for some phenolic resins.

Wedge-Bend-Test, combined with Beckopox EP 307

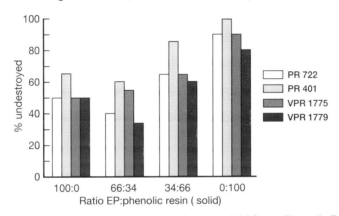

Figure 5 Difference In Flexibility (Fast Deformation) Of Some Phenolic Resins

In Figure 5, another method of measuring flexibility is shown. Four different ratios between a high molecular weight epoxide resin and the appropriate resole have been evaluated: 100:0, 66:34, 34:66 and the pure resole films without any high molecular weight epoxide resin are displayed. The substrate is tinplate (ETP), the film weight approximately 5 grams $\bar{m}^2$ and the stoving cycle 12 minutes at 200°C in an air circulation laboratory oven. A wedge bend tester (fast deformation process) Gardner IG1125 has been used for evaluating flexibility and the figures given above reveal the percentage of the part of the wedge which is destroyed or cracked. To visualise the defects better, the panels under test were immersed in a 10% copper sulphate solution for 10 minutes. A magnifying glass (10x) was used to aid the evaluation. From the results it can be seen that a very reactive resole such as Phenodur PR 401 has less flexibility than a lower reactivity type such as Phenodur PR 722 or Phenodur

VPR 1779. The differences in flexibility between the types VPR 1779/VPR 1776 and PR 722 are less pronounced compared to the results from the cupping test.

(ii) **Substrates**

The substrates discussed below play a major role in the functioning of packaging. They have been developed constantly over the last 50 years and more and have reached a high level of sophistication. Some figures (for some European countries only) are given in Figure 6 to show the importance and in Figure 7 the volumes of metallic substance substrates.

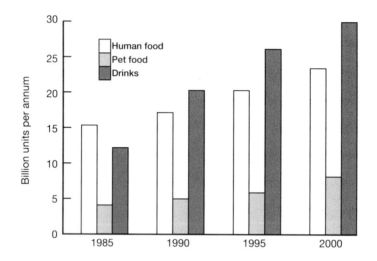

Figure 6 The Estimated Demand for Open Top Cans by End Use.

The metallic substrates used to manufacture containers are always rather thin and consist of either:

- cold rolled steel

- electrolytic tinplate

- tin free steel

- aluminium

(a) *Cold rolled steel (CRS)*
This is mainly used for drums and in some cases for pails. It is frequently pre-treated with phosphates to improve wetting and corrosion protection. Sometimes, CRS is rather difficult to coat, because residues of the pre-treating process are still on the surface.

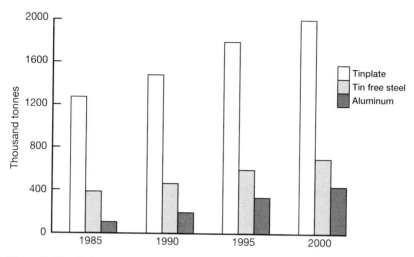

Figure 7 The Estimated Demand on Metal for the number of cans from Figure 6.

(b) *Electrolytically tinplated steel (ETP)*
This is mainly used for three piece cans and general line containers. Tinplate nearly always has some lubricant from its processing on its surface. This lubricant, in many cases paraffin of low molecular weight, diffuses during the stoving process of the lacquer. The tinplate used for laboratory tests should be as fresh as possible ('virgin plate'), because the tin plate surface undergoes a certain degree of degradation during storage. Old tinplate often gives irreproducible results. Tinplate is the only metallic substrate which is effected by sulphur and very special care has to be taken to avoid sulphur staining.

(c) *Tin free steel (TFS)*
Tin free steel has substantially replaced ETP in three piece can ends, since no (expensive) tin is used. The steel base is instead electrolytically coated with a very thin (fraction of a μm) layer of nickel and or chromium. TFS usually gives better results in retort resistance, but is somewhat less satisfactory so far as adhesion properties are concerned. Care must be taken to choose and formulate the right system. However, because of the need for welding three piece can bodies, ETP cannot be replaced by TFS in all cases, because TFS is very difficult, if not impossible, to weld easily. Whilst traditional three piece food can ends are normally made from TFS, the can body is made from ETP.

(d) *Aluminium*
Aluminium is significantly softer than steel and more flexible. It came into this market with the development of two piece, drawn and wall ironed, together with drawn and redrawn cans. In addition to this, easy open ends and ring pull systems require this base metal. There are numerous different aluminium alloys being used, although in the USA, most can

producers use the same alloy to improve recycling. Aluminium is, per se, more expensive than steel but this is counterbalanced partially by reduced weight and better processing properties. Basically aluminium can be coated and the coatings can be cured at a somewhat faster speed than steel. Beyond this recycling of aluminium it is also frequently less expensive. Aluminium is the perfect substrate for coil coating processes and the coating process is frequently done by the aluminium manufacturers themselves. Some of the energy which is set free during the processing of aluminium is used to cure coatings or printing inks in the factory.

Collapsible tubes are still being made from aluminium on a large scale, although, in many cases, tubes made from polyolefines or certain composite materials have replaced them. These plastic tubes have some disadvantages, one of them being the fact that air is inevitably sucked into the tube during usage. This cannot be tolerated by many pharmaceuticals and other air sensitive products like ketchup or mayonnaise. Hence the continuing use of metal tubes for these applications.

(iii) **Containers**

(a) *Two-Piece Cans*
There are basically three different designs of such cans;

- drawn and wall ironed beverage cans (DWI)

- drawn food cans

- drawn and redrawn food cans (DRD).

Drawn and ironed food cans (D&I) are a variant between DWI and drawn cans and here will be considered in the drawn can category. Each of these three can types will be considered separately as the demands placed on the coatings are different for each can making process and type of can produced.

Draw and Wall Ironed Cans (DWI)

These are made either of aluminium or ETP (electrolytic tinplate) and have replaced the older 3 piece design for beer & beverages in all cases except for non-carbonated fruit juice and similar products. In these cases a stronger can is required. The carbonation of beverages allows thinner metal to be used for DWI cans, because the pressurisation imparts rigidity to the can. These cans are of a sophisticated intricate design and are produced at enormously high speeds, 1500–2000 DWI cans per minute being typical. The DWI process begins with coil stock, which is formed into 'cups'. The cup is then lengthened in a wall iron press, where it is pushed through a series of progressively smaller die rings. The top of the can is trimmed to the desired height, and processing lubricants are then removed.

The can is then externally decorated. The interior coating is sprayed into the (nearly) ready made can and it is almost complete and phenolic resins play no significant part in this application. The usual interior lacquer is of a waterborne nature and has to be resistant to pasteurisation only. No sterilisation process is required for beers and beverages. After curing, many cans are now necked to save material cost for the (more expensive) end. The ends used finally to close this sort of cans very often come from coil coated stock and are lacquered separately. This end coating must exhibit a much higher flexibility than the interior spray coating, since the manufacturing process calls for extreme flexibility, high corrosion resistance and non-porosity. Can ends are often coated with PVC containing systems and phenolic resin containing primers are used as adhesion promoters for the organosol lacquer. Curing cycles for such systems are very short indeed, with a typical curing schedule being 20 seconds at 250°C. Coatings for DWI beer & beverage cans are considered in depth in Volume 1 'Acrylics' and Volume II 'Epoxies'.

Drawn Foodcans

Drawn cans are very often small food cans frequently used for packaging fish. They are user friendly with an easy to open ring and pull, or lever opening system. Compared with the older 3 piece can design, there is much less danger of spillage and a much less troublesome opening procedure. One does not need a special opener for this design. This class of single drawn cans are often referred to as 'shallow drawn' or 'club cans'.

The can body is frequently stamped out from coil coated aluminium stock. As always in coil coating, curing cycles are extremely short, ranging in practice from 60″/250°C (PMT) to only 20″/340°C (PMT). PMT is the abbreviation for peak metal temperature and means, that during the given stoving time, the metal surface reaches and holds that temperature for that time. This is confirmed by a coloured thermochromic strip attached to the surface of the metal or by a data pack equipment put through the oven together with the coated panel.

There is a major difference in coating application method between two piece food cans and DWI beverage cans. The interior coatings of drawn food cans are applied before the can making process begins. An exterior coating is applied as well, before mechanical processing, to protect the can during processing. The interior and exterior coating system must withstand the stresses of can formation. It is interesting to know that the external coatings can also be applied before drawing. The so called Distortion Printing Process is used to achieve this and in this a recognisable design appears only after the drawing process is complete. Highly computerised design methods are necessary, together with perfect printing equipment.

Only a few available phenolic resins, mainly combined with epoxide resins of the type '9' can be used for this purpose, some of them are ;

- Phenodur PR 217
- Phenodur PR 612
- Phenodur VPR 1775 and VPR 1776.

Since the deformation takes place at very high speed, it is nearly impossible to correlate results found in the laboratory (and produced by slow speed deformation) with reality. Final tests have to be conducted on the lines themselves to confirm the suitability of a specific system. Formulation 7 gives an example of such an interior coating for drawn food cans.

FORMULATION 7
COIL COATING INTERIOR LACQUER FOR DRAWN FOOD CANS (ALUMINIUM)

MATERIALS	wt.%	solid
Phenodur PR 612/80%BUT	15.50	12.40
Beckopox EP 309	28.80	28.80
Glycol ether	35.00	
High boiling aromatic solvent	19.00	
Additol XK 406(catalyst)	1.20	
PE/PTFE wax	0.50	0.50
TOTAL	**100.00**	**41.70**

Some properties of Formulation 7
Ratio epoxide:phenolic resin: 70:30 (solid resins)
Typical solids content at application: 30%
Typical viscosity DIN/ISO 53211/23°C   130s
Typical stoving schedule (Coil Coating)   25"/320°C(270°C PMT)

The phenolic resin is first pre – diluted with an appropriate part of glycol ether. Solid epoxide resin is dissolved in the remaining part of the glycol ether plus some of the aromatic solvent and both solutions are mixed thoroughly. The catalyst (Additol XK 406) is added to this mixture afterwards, together with the remaining solvent. Additol XK 406 is a phosphoric acid containing curing agent. Compared to pure phosphoric acid, it has significantly better compatibility and therefore has less effect on storage stability.

Waxes are usually dispersed and this can easily be done in a part of the epoxide resin solution. A high speed dissolver or a triple roller mill are suitable equipment for this dispersion process, particularly as some micronised waxes need little dispersion. The phenolic resin used in this case needs some catalyst because it is of medium reactivity, but it has excellent flexibility and for this coil coating process, both properties are of utmost importance. Since the flow properties must be as perfect as possible, a silicon levelling agent might be added, but of course, the food law regulations must be taken into consideration for the additives.

Another way to improve flow and porosity is to pre-condense the phenolic resin with a high molecular weight epoxide resin. The resin mixture is diluted to approximately 40% solids content, and is introduced in the above suggested formulation levels. The mixture is kept under reflux in a suitable kettle (possibly steam heated for a better [faster] temperature control) for a few hours at temperatures between 100 and 130°C. This leads to a slight increase in molecular weight and therefore to a somewhat higher viscosity compared to the cold blend, but this disadvantage is offset by a significant gain in flow and a big improvement in porosity. Highly etherified resoles, which should be not too reactive, are the most suitable for pre-condensation. A limiting factor however is the need for suitable and safe equipment. This, together with reduced application solids, are the main reasons for the decline of the method.

A type '9' epoxide resin could be used instead of the type '7', which would improve flexibility further, mainly with a disadvantageous reduction of solids content at application viscosity (approximately 80s/4mm cup). The Phenodur PR 612 exhibits, under these stoving conditions, a yellow / golden colour. Formulation 8 is for a more reactive system, also for drawn food cans made from aluminium.

FORMULATION 8
COIL COATING INTERIOR LACQUER FOR DRAWN FOOD CANS (ALUMINIUM), VERY HIGH REACTIVITY

MATERIALS	wt.%	solid
Phenodur VPR 1775/70%	18.00	12.60
Epikote 1007	25.00	25.00
Phenodur PR 308/62%	5.00	3.10
Glycol ether	25.00	
High boiling aromatic solvent	25.50	
Additol XK 406 (catalyst)	1.50	
TOTAL	**100.00**	**40.70**

Some properties of Formulation.8	
Ratio epoxide:phenolic resin:	70:30 (solid resins)
Typical solid content at application:	35%
Typical viscosity DIN/ISO 53211/23°C	120s
Typical stoving schedule (Coil Coating)	15"/335°C(280°CPMT)

The phenolic resin Phenodur VPR 1775 is a lightly coloured, very reactive butylated resole, containing practically no free monomers (phenol / cresole). During application, there is no detectable 'phenolic smell'. The resin is first prediluted with an appropriate part of the glycol ether. The solid epoxide resin is dissolved in the remaining part of the glycol ether plus some of the aromatic solvent and both solutions are thoroughly mixed. Phenodur PR 308 is a colouring resin used to tint this lacquer. The catalyst (Additol XK 406) is added afterwards, together with the remaining solvent. The phenolic resin used in this case is not very reactive and therefore needs a catalyst. Since flow properties must be as perfect as possible, a silicon levelling agent might be added as it was before. Again a type '7' epoxide resin could be used instead of a '9' type. Formulation 9 describes a coil coating for can coating systems with a higher reactivity, compared to Formulation 8.

FORMULATION 9
COIL COATING INTERIOR LACQUER FOR ALUMINIUM AND TFS

MATERIALS	wt.%	solid
Phenodur PR 217/65%BUT	17.00	11.05
Beckopox EP 309	25.00	25.00
Maprenal MF 800/55%BUT	2.00	1.10
Glycol ether	25.50	
High boiling aromatic solvent	25.50	
Additol XK 406(catalyst)	1.50	
Lubricant solution	3.50	
TOTAL	**100.00**	**37.15**

Some properties of Formulation III.9	
Ratio epoxide:phenolic resin:	70:30 (solid resins)
Typical solid content at application:	30%
Typical viscosity DIN/ISO 53211/23°C	80s
Typical stoving schedule (Coil Coating)	90"/261°C(230°CPMT)

Formulation 9 is preferably used in coil coating systems, which are not cured at the very highest temperatures. The phenolic resin Phenodur PR 217 is a medium coloured, reactive resole. The resin is first pre-diluted with an appropriate part of glycol ether. The solid epoxide resin is dissolved in the remaining part of the glycol ether plus some of the aromatic solvent and both solutions are thoroughly mixed. ®Maprenal MF 800 is a butylated melamine / formaldehyde resin and there are three main reasons for adding such a material, namely;

- The melamine resin improves reactivity

- The melamine resin improves flow, thereby reducing the degree of porosity

- The melamine resin increases the solid content

The catalyst (Additol(XK 406) is added to this mixture afterwards, together with the remaining solvent(s). The phenolic resin used in this case needs a catalyst, and the ®Additol XK 406 improves, at the same time, the adhesion of this lacquer. The lubricant solution (for example, a solution of lanolin wax) is added as the last part of manufacture into the lacquer.

Drawn and Drawn/Redrawn (DRD)

Drawn and drawn redrawn cans have to undergo more than one step of deformation. The first draw cannot produce a can which is deep enough, so that a second and even a third deformation step may follow. This type of can is also called a 'deep drawn can'. Contrary to the DWI can, the wall thickness of DRD cans is not reduced during processing. Some of these cans are made of pre coated stock, either coming from coil or sheet coating processes. The phenolic resins used for this application must necessarily exhibit a rather high intrinsic flexibility and the number of suitable grades is limited. Exterior coatings can only be applied after the can has been drawn and in some cases, a dip coat is applied externally as the last coating procedure. This clear dip coat protects the often external graphic design of the can. Suitable phenolic resins for this sort of application are, for example;

- Phenodur PR 722
- Phenodur PR 897
- Phenodur VPM 1777;

In this application, these resoles are usually co-curing resins for type '9' epoxide resins. Phenodur VPM 1777 is however not a resole, but a carboxyl functional novolac, and via this functionality, allows more direct curing with the epoxide resin without having the tendency for self-condensation. This leads to a very high flexibility of the cured coating system. The other two resoles have more conventional chemistry, leading to reactive and flexible (for PR 897 even more flexible) coatings with good retort and chemical resistance. Formulation 10 gives an example of such an interior coating system.

Phenodur PR 722 is a reactive resole, exhibiting very good retort resistance in addition to good sulphur staining properties. Phenodur PR 308 is used as a colouring resin to achieve the

FORMULATION 10
INTERIOR GOLD LACQUER FOR DRD CANS

MATERIALS	wt.%	solid
Phenodur PR 722/53%BG/BUT	15.00	7.95
Type '9' epoxide resin	25.00	25.00
Phenodur PR 308/62%	5.00	3.10
Glycol ether	25.00	
High boiling aromatic solvent	25.50	
Additol XK 406(catalyst)	1.50	
Lanoline wax solution	3.00	
TOTAL	**100.00**	**35.95**

Some properties of Formulation 10	
Ratio epoxide:phenolic resin:	70:30 (solid resins)
Typical solid content at application:	35%
Typical viscosity DIN/ISO 53211/23°C	80s
Typical stoving schedule (Coil Coating)	60s. at 250°C (PMT)

(often) required 'gold' colour shade. Additol XK 406 is the catalyst used for this coil coating application. The Lanolin wax improves the smoothness of operation during the processing of the coil.

In the USA mainly, there is a 'mixed design' between DWI and DRD which is quite common. This is the drawn and ironed food can (D&I). In such cases, the interior coating has to be applied by a spray application and the external system may be a dip coat (after the metal offset printing process). Many pet food and soup cans sold in the USA are D&I cans. This type of can is becoming more popular in Europe. However, due to the capital investment cost and the costs of changing tools for a differently styled can, the DI process is only viable for the most popular, standard sized cans.

(b) *Three-Piece Cans*

This is the oldest design model for a food can. It consists of the body itself and two ends, which together make three pieces which form the complete can. The body is a rectangular piece of electroplated tin (ETP) to which is rolled to form cylinders. Three piece cans are nearly always made from sheet metal. The sheet is coated with the internal lacquer first. Some areas are left uncoated (they are termed reserves) to facilitate the welding. Three piece cans are made by basically 3 processes:

– Soldering

This is the oldest process, in which the edges of the blank are notched and folded together. This hook like joint is pressed together first mechanically and then solidly by the use of flux and solder. Due to the health risk from the solder (lead), this technology is no longer used to weld cans for wet food stuff. In fact, people consuming this type of packed foodstuff for prolonged periods of time have frequently been found to suffer from lead poisoning.

– Cement Process

In this process adhesives are employed instead of solder, but this process is rarely used in practice.

– Welding

This is the most sophisticated process and is extensively used to produce three piece cans. The body of the can is formed by overlapping joining surfaces, then tack welded, followed by wheel or wire welding.

In every case, the seam which is produced in this way has to be coated to protect both the metal and contents of the can from any interaction between them. This is frequently done by side seam striping systems, such as powder coatings based on thermoplastic polyester powders or by the means of special liquid coatings containing high molecular weight epoxide resins, phenolic resins or epoxide urea systems. The ends of the cans are normally stamped from coated tin free steel (TFS). Due to the abrasive natute of this substrate, TFS ends are always coated before stamping to prolong the life of the press tools. One end is attached so that the can is ready to be filled, before the other end is sealed under vacuum for further processing.

The coating process of the flat substrate is often called a sheet fed process, where individual sheets are fed into a roller coater and subsequently into the oven. A typical stoving cycle is 12 minutes at 200°C, thereafter the sheets are often cooled by a stream of cold air and stacked together for further processing. In some cases, the exterior coating is applied after the interior coating and care must be taken, to ensure that the interior coating is not spoiled by an overflow of the exterior system or by effluxes coming from the exterior coating during the second or subsequent stoving cycle. The following Formulation, 11 may be considered as a 'standard' interior system for a three-piece can.

FORMULATION 11
STANDARD INTERIOR SYSTEM FOR 3-PIECE CANS

MATERIALS	wt.%	solid
Phenodur PR 897/53%BG/BUT	15.00	7.95
Type '7' epoxide resin	25.00	25.00
Phenodur PR 308/62%	5.00	3.10
Glycol ether	25.00	
High boiling aromatic solvent	25.50	
Additol XK 406(catalyst)	1.50	
Lanoline wax solution	3.00	
TOTAL	**100.00**	**35.95**

Some properties of Formulation 11	
Ratio epoxide:phenolic resin:	70:30 (solid resins)
Typical solid content at application:	35%
Typical viscosity DIN/ISO 53211/23°C	65–70s
Typical stoving schedule	12 min. at 200°C

The resole used in this formulation introduces a very high degree of flexibility, a light colour and excellent chemical resistance. Since it is of relatively low reactivity, the use of a catalyst is recommended. The high molecular weight epoxide resin is first pre-diluted with the glycol ether, the resole added to it and all the other ingredients added in the order given above. Phenodur PR 308 is used as the colouring resin to darken the relatively light colour of the resole. Cured at sheet fed cycles like 12 minutes at 200°C, the lacquer shows excellent flexibility and retort resistance, for example, against 2% lactic acid or a mixture of 2% acetic acid and 3% potassium chloride.

The phenolic resin used in this formulation demonstrates that reactivity or cure measured by the degree of solvent resistance (methyl ethyl ketone or acetone double rubs) is not always directly linked to chemical resistance. Solvent resistance as a measurement of the degree of crosslinking is, or is not, related to chemical (retort) resistance (Figure 8).

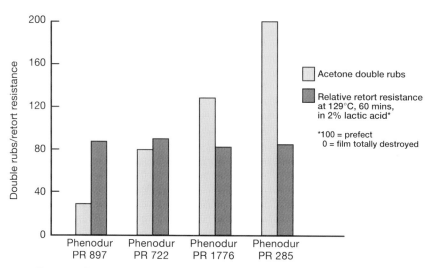

Figure 8 Solvent Resistance versus Retort Resistance of some Resoles

(iv) Drums And Pails

The exterior and interior coating of drums and pails still remains a very important part of the container coating industry. Normal thermosetting alkyd / melamine resin systems are mainly used for exterior coating. Phenolic resins are used often as the sole binder or as the main resin in these formulations. The interior coating of drums must withstand much more aggressive goods or chemicals compared to interior food can coatings. Resistance is measured by immersing the coated test panels into the test substance and by storing at elevated temperatures (say: 200 days at 40°C). A typical list of resistance properties is given in Table 5.

Because of the extreme degree of chemical resistance required, it is necessary thoroughly to cure the interior coating. To achieve this, the use of flexibilising resins such as epoxide resins or poly vinyl butyrals must be limited. The interior coatings of drums, therefore, only contain the most chemically resistant resoles and only small amounts of other flexible resins. The ratio between phenolic resin and flexibilising resin may be 60–95 parts of phenolic resin to 40-45 parts of epoxide resin or poly vinyl butyral.

TABLE 5 : LIST OF SOME SUBSTANCES, AGAINST WHICH INTERIOR DRUM LININGS MUST BE RESISTANT

Chemicals:	
Hydrochloric Acid 30%	Sodium Chloride
Hydrochloric Acid 20%	Sodium Hydroxyde
Acetic Acid	Ammonia
Acetic Acid	Synthetic Detergent Solutions
Acetic Acid	Soap
Citric Acid	Resins
Lactic Acid	Paints (Emulsion Paints Often Contain Ammonia!)
Solvents:	
Ethanol	Ethanol
Butanol	Turpentine
Methanol	Acetone
Xylene	Methyl Ethyl Ketone
Benzene	Methoxy Propanol
Toluene	Methoxy Propyl Acetate
White Spirit	
Food	
Mustard	Orange Juice
Ketchup	Mayonnaise
White Wine	Beer
Pineapple Juice	Lard
Tomato Juice	Olive And Other Vegetable Oils

The substrate of which most drums are made is cold rolled steel, often used without pre-treatment and frequently with residues of grease, lubricant or even corrosion products adhering to the surface. Pails on the other hand are mainly made of ETP and this surface is less difficult to coat than cold rolled steel.

Interior coatings for drums are usually pigmented, mostly by iron oxide. This and other inorganic pigments must have soluble salt levels which are as low as possible, otherwise problems in chemical or water resistance occur.

For pails (used frequently for liquid coatings, emulsion paints and similar products), unpigmented interior coatings are standard and the raw material for pails is often ETP sheet. The pail is formed after the sheet has been coated.

Since interior coatings for drums are mainly applied after the drums have been formed, only a limited degree of flexibility is required and this, in turn, makes it possible to use coatings based upon resoles as the main or sole binder. These coatings are applied by spray and in some cases, where maximum chemical resistance is required, drums are coated (and stoved) twice. A typical stoving schedule for a drum and pail interior coating is 15 minutes at 230°C. Three formulations, one for a clear and two for pigmented interior coating, for drums are:

FORMULATION 12 : CLEAR PAIL COATING

MATERIALS	wt.%	solid
Phenodur PR 373/53BG/B	55.00	29.15
Butanol	2.90	
Mowital B 30T	5.00	5.00
Butanol	22.00	
Butyl glycol	6.00	
Methoxy propanol	7.10	
Solvesso 100	2.00	
TOTAL	**100.00**	**34.15**

Some properties of Formulation 12	
Ratio: phenolic resin: PVB	90:10 (solid resins)
Typical solid content at application:	25–28 %
Typical viscosity DIN/ISO 53211/23°C	20-30s(at application)
Typical stoving schedule	10 min. at 260°C

Phenodur PR 373 is a very reactive phenol resole, with little etherification. It has limited compatibility with high molecular weight epoxide resins. In this formulation, the resole is combined with poly vinyl butyral (PVB) to achieve a certain degree of flexibility. The Mowital grade B 30T is a poly vinyl butyral of medium viscosity, containing a relatively high degree of hydroxyl functionality. This leads to a higher degree of crosslinking between the resole and the PVB. The poly vinyl butyral powder is dissolved in butanol and added to the phenolic resin. The lacquer is completed by adding all other ingredients subsequently. No catalyst is necessary, firstly because the resole used is of high reactivity and secondly stoving temperatures are comparatively high. Last, but not least, the catalyst may reduce chemical resistance (acid catalyst versus alkaline resistance!). This type of lacquer has a shelf life of 4-6 months at room temperature. Formulation 13 is for a red pigmented interior drum liner.

Formulation 13 is, with the exception of the inorganic pigment used, similar to formulation 12. The poly vinyl butyral powder is dissolved in butanol. Iron oxide is dispersed into the resin by bead mill (0.5 to 1 mm beads) at approximately $20 \mathrm{ms}^{-1}$ circumference speed. In volume terms, the amount of beads used should equal the amount of the resin pigment pre-mix. The resole used in this formulation is a very reactive one, therefore care should be taken that the temperature during grinding does not exceed 40°C, otherwise there might be self-condensation reactions, leading to viscosity increase. The ball mill is washed with the PVB solution, followed by the other solvents. The fineness of grind should be carefully controlled

FORMULATION 13 : RED INTERIOR DRUM LINING

MATERIAL	wt.%	solid
Phenodur PR 373/53BG/B	50.00	26.50
Butanol	2.90	
Mowital B 30T	5.00	5.00
Butanol	12.00	
Iron oxide red 510	15.00	15.00
Methoxy propanol	3.10	
Solvesso 100	2.00	
TOTAL	**100.00**	**34.15**

Some properties of Formulation 13	
Ratio: phenolic resin: PVB	90:10 (solid resins)
Typical solid content at application:	30-32 %
Typical viscosity DIN/ISO 53211/23°C	20-30s(at application)
Typical stoving schedule	10 min. at 260°C

FORMULATION 14 : RED DRUM LINING

MATERIALS	wt.%	solid
1 Phenodur PR 260/68BG/B	50.00	34.00
2 Butanol	2.90	
3 Butanol	12.00	
4 Iron oxide red 510	15.00	15.00
5 Methoxy propanol	3.10	
6 Solvesso 100	2.00	
TOTAL	**100.00**	**49.00**

Some properties of Formulation 14	
Typical solid content at application:	35 %
Typical viscosity DIN/ISO 53211/23°C	20-30s(at application)
Typical stoving schedule	10 min. at 260°C

and must be below 10 microns to guarantee a perfect, non porous film with a good flow. For the same reason, filtration is also important.

Chemical resistance in this coating is high, based on the resole. Iron oxide red is dispersed by a bead mill (0.5 to 1mm beads) in phenolic resin at approximately $20m^{s-1}$ circumference speed. To achieve low grind viscosity, the resole is pre-diluted with approximately 50% of butanol in the formulation. The amount of beads used should equal the amount of the resin pigment pre-mix. The ball mill is washed with remaining solvents. Grind must be below 10 microns to guarantee a perfect, non porous film with a good flow. For the same reason, filtration is also important. The coating has little flexibility but good resistance, even against aggressive chemicals such as ammonia solution.

Coatings of this type may also be used to coat equipment like kettles, pipes, heat exchangers or other apparatus used, for example, in the chemical industry, or in power plants. This type of equipment is often coated as many as eight times to reduce porosity to nil to obtain maximum resistance. To prevent cracking of this multi layer system, due to internal stresses, the first coats are not fully cured but are only 'partially dried'. The full cure takes place only after the last coat is applied and a typical schedule is 20 minutes at 160°C for the first coats and 30 minutes at 270°C for the last.

(v)   Collapsible Tubes

Collapsible tubes offer ingenious, and easy packaging for pastes and other materials with similar liquid viscosities and flow characteristics. Some examples of materials packed into collapsible tubes inlcude:

TABLE 6 : SUBSTANCES PACKED FREQUENTLY IN COLLAPSIBLE TUBES

Mustard	Medicines
Mayonnaise	Ointments
Ketchup	Adhesives
Fish And Caviar Paste	Inks
Shaving Cream	Colours
Onion & Garlic Paste	

Some, if not all of these products are frequently aggressive. Medicines may well be sensitive to oxygen, and contain highly active substances. A substantial part of the food packed into collapsible tubes is highly acidic, highly coloured and very sensitive to oxygen. Up to the early sixties, collapsible tubes were manufactured from tin. Since this is a rare metal and comparatively expensive, the industry changed to aluminium. More recently, plastic tubes became very common and captured a large share of the market from aluminium tubes. However, some very sensitive or very aggressive substances are still packed in metal tubes and will keep a portion of the market for the foreseeable future. One major disadvantage of plastic tubes is, that after pressing and releasing the pressure, air is sucked back into the tube. The given shelf life of food or medicine packed in collapsible tubes is, on the average, similar to that of food packed in two or three piece cans and interior coating must be as chemically resistant.

The manufacturing process for collapsible tubes starts with an aluminium pellet which is expanded in a complicated multi step process to become a tube, open at one end and with a thread at the other. The thread is generally closed by a thin membrane, also made of aluminium and formed directly as the tube is manufactured. Line speed is not very high, compared with DWI can making, however, one line might be able to produce as many as 400 tubes per minute.

The interior coating is always applied by spray, so that lacquers are sprayed into the tubes with a spray gun (lancing) with a long nozzle. Either the tube or the nozzle, or even both turn during the spray process, so that an even lacquer distribution is assured. After spraying, the tubes are put into small baskets, one for each tube. The baskets are mounted onto an endless chain which is moving constantly forward through a short flash off zone and into the oven. The curing time is much shorter than that for three piece can interior systems, typically being 5 minutes at 250°C. After the curing process, the tubes are coated externally, usually with a white or coloured basecoat, the printing ink and finally a clear overprint varnish.

Some tubes seen on the market have excellent graphics. Both, the exterior and the interior coating system must have extremely good flexibility, since collapsible tubes are the only packaging material, which is deformed during its use. The interior coating must achieve a balance between chemical resistance and flexibility. This is achieved by using a high curing temperature, a high molecular weight epoxide resin and a highly resistant resole. In most cases the interior coating must be formulated in accordance with relevant food laws. The following formulations should be a guide to formulation of interior coatings for collapsible tubes.

FORMULATION 15
HIGH REACTIVITY INTERIOR COATING FOR COLLAPSIBLE TUBES

MATERIALS	wt.%	solid
Phenodur PR 285/55%	13.00	7.15
Beckopox EP 307/100%	27.00	27.00
Methoxy propanol	27.00	
Maprenal MF 800/55%	1.40	0.77
Solvesso 150	15.00	
Ethoxy propanol	16.60	
TOTAL	**100.00**	**34.92**

Some properties of Formulation 15
Viscosity DIN 53211/23°C/4mm   : approx. 80"
Typical stoving cycle   : 4 min./260°C
Spray-diluent : Butyl acetate/Methoxy propanol/Xylene: l 1:1:1
Ratio epoxide:phenolic resin   : approx. 80:20 (solid)

Here again, a highly reactive resole is used as the co-curing agent for a high molecular weight epoxide resin. The solid epoxide is first diluted in methoxy propanol, before all the other solvents, the resole and the melamine resin are added. Maprenal MF 800 is a medium reactivity butylated melamine resin, used in this case, at a low level, as a flow agent to optimise the wetting of the lacquer onto the aluminium. A high boiling solvent ethoxy propanol is used in this formulation, to cope with the problem of a short flash off time combined with high initial oven temperatures.

FORMULATION 16
INTERIOR COATING FOR COLLAPSIBLE TUBES, DARK "GOLDEN" COLOUR

MATERIALS	p.b.w.	solid
Phenodur PR 723 / 60%	15.00	9.00
Beckpox EP 307/50%diluted in Methoxypropanol	53.60	26.80
Maprenal MF 590/55%	0.50	0.30
Additol XK 406 (catalyst)	1.20	
Methoxy propanol	10.50	
Diacetone alcohol	12.40	
Solvesso 150	6.80	
TOTAL	**100.00**	**36.10**

Some properties of formula 16	
Viscosity DIN 53211/4mm/23°C	:approx. 90"
Ratio epoxide : phenolic resin	:approx. 25:75
typical cure cycle	:3-5 min./240-290°C

(vi)   **Aerosol Cans**

Aerosol cans are rather costly, so that generally only relatively expensive goods are packed in them. Great efforts are made to coat the exterior of such cans to as high a quality as possible, to enhance the image of the (expensive) content. Aerosol cans are emptied by an intricately constructed valve and always contain pressurised gas (or highly volatile liquid) which is responsible for the transport of the contents. A similar design, but without the valve construction, is fitted with a thread and used mainly for refreshing drinks.

The interior coating must not only resist the contents of the can, but also the propellant, which might be, for example butane, halogenated hydrocarbons or dimethyl ether. These propellants are aggressive solvents and the can is pressurised. For some contents, such as mineral oils, it may not be necessary to use an internal lacquer, this being a supplier's decision. Failure of the aerosol internal coating, however, could possibly result in an explosion, due to the perforation of a pressurised can. Interior coatings, moreover, are often formulated according to food laws, since some, but not all, products filled are food or food related products. Possible contents for aerosol cans are given in Table 7.

TABLE 7 : SOME POSSIBLE CONTENTS OF AEROSOL CANS

Medicines	Shoe Cleaner
Hair Spray	Oven Cleaner
Deodorants	Cream
De-Icing	Adhesives
Solvents	Lacquers
Perfumes	Paints

Aerosol cans are manufactured by two different routes. In essence, they mirror the 2 and 3 piece cans, if the domed end of an aerosol can is considered as a can end. Internal lacquers are predominantly epoxy/phenolic based, (over 65% of the market), with polyimines accounting for about 5% of the market for the most demanding products packed. The remainder are normally based upon PVC containing coatings.

Traditional three piece aerosol cans are made from a body with a bottom end seamed on. Three piece aerosols are made from tinplate (ETP). The epoxy phenolic internal lacquer may be roller coated onto the sheet with an edge (reserve) left on the blank for welding. In some cases, 3 piece aerosol cans may be internally sprayed, in addition to lacquering as a sheet, after the body has been formed and the end seamed on, for additional protection.

The equivalent of the 2 piece can is the "monobloc" aerosol can. These cans represent a relatively new packaging system. The cans are made of aluminium and are drawn in a multi step process from a big aluminium tablet. Monobloc aerosols are more popular in Europe than North America. The internal coating is often a PVC based coating, spray applied by a long gun (lancing). The production rate is much slower than that for DWI beverage cans, being at most a few hundred per minute.

Exterior coatings are usually applied by roller coating, and in some cases, a clear overprint varnish is applied by spray application.

The interior coating is always applied by spray application at a relatively high film thickness, of the order of 10–20g per m^2. After a relatively long flash off time (because of the small opening), the cans are stoved usually for 10–20 minutes at 230°C. Formulation 17 is typical for an interior lacquer for aerosol cans.

Formulation 17 contains a reactive resole (phenol / cresole type, which is combined in a ratio of approx. 30:70 (solid / solid) with a high molecular weight epoxide resin, Epikote 1007. A

FORMULATION 17 : INTERNAL SPRAY LACQUER FOR AEROSOL CANS

MATERIALS	wt.%	solid
Phenodur PR 722/53BG/B	18.90	10.00
Epikote 10007/50%MP	44.00	22.00
Butyl acetate	10.50	
Di acetone alcohol	10.00	
Methoxy propanol	6.00	
Solvesso 150	9.10	
Urea resin 60B	1.50	0.90
TOTAL	**100.00**	**33.90**
Some properties of Formulation 17		
Typical solid content at application:	33 %	
Typical viscosity DIN/ISO 53211/23°C	20-30s(at application)	
Typical stoving schedule	20 min. at 220°C	

small amount of a reactive urea resin is added to increase crosslink density and to improve flow and adhesion on the aluminium. The solvent mixture given in this formulation is adjusted in such a way, that a portion of it evaporates quickly (butyl acetate), so that during the flash-off period, most of it can leave the film and exit the small opening of the aerosol can. Due to the viscosity increase of the lacquer, this in turn prevents sagging. Other solvents used evaporate less quickly to ensure good flow and levelling. The resole exhibits, after stoving, a dense gold tone. The difference between the can coating stoving cycle of only 200°C and that for internal coating systems for aerosols is very clear. Formulation 18 is for an interior spray lacquer for aerosol cans with a very high solvent resistance.

This formulation is based on a strongly reactive resole and has excellent solvent resistance properties. The high molecular weight epoxide used is diluted in methoxy propanol first, before all the other ingredients are added. This formulation contains no catalyst, since the resole itself is very reactive already. The viscosity of the lacquer is relatively high and must be reduced before spraying, using the above mentioned spray diluent. The colour of the cured film is medium gold. The melamine resin used is a highly reactive, butylated one. The mela-mine resin also improves flow and wetting. Such a lacquer is well suited to coat the interior of aerosol cans used to spray lacquers, coatings and related products (for example, thermoplastic acrylic resin based automotive repair systems and do-it-yourself spray coatings).

Relatively new in this industry (at least in Europe) is the growing need for waterborne systems. As a basic rule, waterborne systems are better suited to spraying than roller coater applications. Nearly always the non-newtonian behaviour of waterborne systems is the main reason for this. Formulation 19 is for a waterborne spray interior coating, with a high degree of chemical resistance.

FORMULATION 18 : INTERNAL SPRAY LACQUER FOR AEROSOL CANS, VERY GOOD SOLVENT RESISTANCE

MATERIALS	wt.%	solid
1 Phenodur PR 285/55%	20.00	11.00
2 Beckopox EP 307/100%	34.00	34.00
3 Maprenal MF 577/55%	10.00	5.00
4 Methoxy propanol	34.00	
6 Solvesso 150	2.00	
TOTAL	**100.00**	**50.00**

Some properties of Formulation 18
Typical solid content at application: 28–30 %
Typical viscosity DIN/ISO 53211/23°C   20–30s(at application)
Typical stoving schedule   10 min. at 200°C
Diluent mixture 30% xylene, 40% butanol, 30% butyl acetate

FORMULATION 19 : WATERBORNE INTERIOR SPRAY COATING

MATERIALS	wt.%	solid
1 Phenodur VPW 1942/56WA	75.00	42.00
2 Hexyl glycol	6.00	
3 Butyl glycol	7.00	
4 Water deionized	9.00	
5. Maprenal MF 904	2.00	2.00
TOTAL	**100.00**	**43.00**

Some properties of Formulation 19
Volatile organic content   approx. 15% on weight
Typical solid content at application: 40 %
Typical viscosity DIN/ISO 53211/23°C   20–30s(at application)
Typical stoving schedule   15 min. at 230°C

Formulation 19 is for a waterborne, sprayable interior can coating system. Phenodur VPW 1942 is an anionic stabilised waterborne phenolic epoxide pre-condensate. It has a very small particle size of less than 100 nm. It is self-crosslinking and contains a certain amount of phenolic resin together with a high molecular weight epoxide resin. The hexa methoxy methyl melamine resin, Maprenal MF 904, is used as flow agent and improves film formation, and reduces 'boiling' tendencies. The hexyl glycol and the butyl glycol are used as film forming and flow / wetting promotion solvents. The solvents also prevent drying of the lacquer at the spray gun nozzle. The solvents have to be added to the resin emulsion with rigorous agitation. The lacquer shows excellent storage stability and resistance properties. Under the right stoving conditions, this waterborne phenolic epoxide gives an intense gold tone.

(vii)   **Closures**

Caps and closures have been used for many years as a sealing mechanism, mainly for glass jars, bottles and containers. In addition to sealing the contents, after filling, they must also to be opened as easily as possible whilst protecting the contents for further storage.

Caps and closures are mainly made of tinplated steel, although those used for pilfer-proof spirit bottles and similar applications are made from aluminium. The deformation takes place after coating, thus both the interior and the exterior coating system must have adequate flexibility. The exterior coating system has to have good optical properties and often consists of a white basecoat, printing ink (s) and a clear overprint varnish. The contents of glass jars can be extremely aggressive. An example is the packing of pickled onions which are treated with sulphur dioxide for storage protection against microbiological attack or putrification. Good corrosion resistance is required as often as not.

Closures mean different things to different people. Jam jar lids, for example, are different from bottle tops or crowns. They are, obviously, however, both closures. Both have a compound or gasket applied to the surface in contact with the glass jar or bottle, which acts as an airtight seal. For jars, this compound is almost always a PVC based plastisol. In the case of sterilised jars, the thread is actually "cut" into this compound. For many bottle tops, PVC has been superseded by alternative materials. All metal internal surfaces have an internal lacquer as a primer, this is normally epoxy phenolic based. The primer is sometimes applied in a coil coating process and is often at a very thin film thickness (not more than 2 gms per m^2) to achieve flexibility for subsequent deformations. In the case of closures for bottles, in contrast to those for jars, a compound may be applied directly to the primer. This primer or basecoat imparts chemical resistance to the closure in the event of the top coat (or sealing compound, in the case of bottle tops) being mechanically damaged during processing or transport to the filling operation.

For closures for jars, a PVC based top coat may be applied to promote adhesion between the PVC compound and the closure. This also improves resistance properties and forms a barrier against the product packed. The basecoat or primer also has the function of an adhesion promoter, because most coatings based on poly vinyl chloride polymers do not adhere well if applied directly onto tinplate, tin free steel or aluminium. This is exaggerated for PVC based compounds or plastisols, which often contain plasticizers which may reduce adhesion further. The sealing materials (often called gaskets) are applied in thick films and must be very elastic to allow manifold trouble free closing and opening operation.

Formulation 20 shows such a base lacquer or adhesion promoter used beneath a sealing compound based on PVC in caps, closures or crown corks.

FORMULATION 20 : ADHESION PRIMER FOR PVC COMPOUNDS IN CAPS, CLOSURES OR CROWN CORKS

MATERIALS	p.b.w.	solid
Phenodur PR 722/53%	45.00	23.85
Plexigum M 527/25%MP/DAA/Sol.150 (1:1:1)(Röhm)	40.60	10.15
Methoxy propanol	3.00	
Ethoxy propyl acetate	4.00	
Solvesso 150	1.24	
Estabex 2307 (stabiliser, OXYDO)	6.16	6.16
TOTAL	**100.00**	**40.16**

Some properties of Formulation 20	
Typical solid content at application:	25–30 %
Typical viscosity DIN/ISO 53211/23°C	50–60s(at application)
Typical stoving schedule	5 min. at 200°C

This unusual mixture of resole and acrylic resin is used to reach a compromise between the required resistance of the base lacquer and the adhesion promotion between the PVC containing sealing compound and the base metal used. Firstly, the thermoplastic, solid supplied acrylic resin is dissolved in a mixture of methoxy propanol, di acetone alcohol and Solvesso 150 (ratio 1:1:1). This solution is then mixed with phenolic resin and stabiliser. Often an epoxidised natural oil, such as soya bean oil, is then added. In this case, the reactive resole Phenodur PR 722 is not combined with a high molecular weight epoxide resin, but with a non – reactive, thermoplastic acrylic resin. This combination ensures good chemical resistance (from the resole), excellent inter adhesion properties (from the acrylic resin) and fast cure or set, since one third of the solid content is a thermoplastic resin

4. Test Methods

(a) *Introduction*
The test methods used to evaluate interior can coating systems are very important indeed . Unfortunately, there are no ISO / DIN / ASTM or BS standards yet, specially related to the requirements of the can coating industry. In some cases, such as adhesion (cross cut) and flexibility (impact tester) standardised test methods might be used, but in many other cases such as retort resistance or cupping , no standards are available. Another specific problem is that, in practice, a filled food can must have a storage stability for a number of years and it is highly impractical to conduct laboratory tests for such a long period of time. Therefore test methods have been established which give some sort of relationship between R & D short cycle tests and practice.

The test methods used should:

- Be practical and as near as possible to the end use

- Be able to give quick and reliable results

- Relate short term tests to long term 'reality'

Some of these test methods are discussed below.

(b)   *Retort Resistance*

Retort resistance is one of the most important properties which a can coating system must possess. Canned human and animal food must not contain preservatives and this is especially true for canned food in 2 and 3 piece cans. Compared with beer & beverages, most wet foodstuff must be sterilised to achieve in-can stability. Most human and animal foodstuffs are acidic (pH ranges from 1 to 6 but some foods are distinctively alkaline and others contain significant amounts of organic bound sulphur derived mainly from amino acids (meat, fish). Salt, used as one of the few permitted preservatives or coming from the food itself, all kinds of organic acids (for example, citric, tannic or oxalic acid) and numerous other naturally occurring chemicals, add to the complexity of the composition of the foodstuff. Since cans are being produced and used not least because of their extremely long shelf life, the testing of retort resistance is a painstaking, and essentially comprehensive undertaking. It is, however, highly impractical to use real foodstuffs as test materials in the application laboratory. This must be, obviously, left to the end user (cannery) working in conjunction with the canmaker. In some cases, interior systems based on phenolic epoxide resins are additionally used on the outside of the container. In these cases, the coating must withstand the sterilisation process, and the water used as its steam source. This is however never straightforward distilled water. Solutions of organic and/or inorganic salts are used to prevent corrosion in the steam container system. The water is always of high pH (approximately 9-10) and for this reason, in practice, food simulants are used for testing can coatings.

(c)   *Food Simulants*

Examples of typical food simulants used to test retort resistance (% on weight) are

Lactic acid test:
2% lactic acid diluted in distilled water

Potassium chloride / acetic acid test:
2% acetic acid + 3% sodium chloride diluted in distilled water

Bouillon – Test
3% Laurylsulfate – Bouillon (supplier Merck) in deionised water + 2 % acetic acid

Sulphur staining test (Fraunhoferinstitute University of Munich)

1. Phosphate buffer solution:

3,56g Sodiumdihydrogenphosphate Merck Art. Nr. 4873
7,22g Disodiumhydrogenphosphate-2-hydrate Merck Art. Nr. 6580 per litre of distilled water.

2. Per 1 l buffer solution 0,5g Cysteiniumchloride Monohydrate Merck Art. Nr. 2839

The solution must be used immediately after preparation. Retort for 90 minutes at 121°C, cool down and store the closed container for another 24 hours.

> Rating:
> 0 = no discolouration
> 1 = very little discolouration
> 2 = little discolouration
> 3 = medium discolouration
> 4 = strong discolouration
> 5 = extremely strong discolouration (marble effects)

Retort water test (for exterior coatings of gold lacquers):
3,56g Sodiumdihydrogenphosphate Merck Art. Nr. 4873
7,22g Disodiumhydrogenphosphat-2-hydrate Merck Art. Nr. 6580 per litre of distilled water.

Simple Sulphur staining test
Test pieces are packed between commercially available white beans. Treatment, test conditions and rating are similar to the sulphur staining test outlined above.

The beans can only be used once. Alternatively, other real foods can be packed and evaluated.

(i) Procedures

Under normal circumstances, correctly coated and cured sheets are cut to a size which fits comfortably into glass or stainless steel containers, which are later introduced to the retort equipment. The cut portions of metal should be immersed into the food simulant in such a way, that a small part of the coated substrate is not immersed. This is helpful later on to check the difference between immersed and none immersed parts of the substrates. If (and this will probably always be the case) more than one panel is tested at the same time, care must be taken, that the parts are separated from each other by means of glass rods or the like. For deep drawn test pieces, the same criteria apply and even more care must be used to separate the test pieces.

(ii) Acid Retort Resistance

The containers containing appropriate test solution (for example, 2% lactic acid) and the test panels are placed in the retort equipment which itself contains the necessary amount of distilled water to form steam. The equipment is closed, heated to the retort temperature and held at an agreed temperature and time. Some practical time / temperature relations are given below:

- 60 minutes at 121°C (approximately 1.4 bar)

- 90 minutes at 121°C (approximately 1.4 bar), standard for sulphur staining

- 60 minutes at 129°C (approximately 1.8 bar), standard in most evaluation laboratories

- 90 minutes at 129°C (approximately 1.8 bar)

(iii) **Blushing**

Blushing is the phenomenon which takes place when water is absorbed into the coating during the retort process. This effect is often not apparent immediately but rather after about one hour. It is a whitish milky shade and is easier to recognise with gold lacquers than with clear coats. Blushing must not be mistaken as a residue coming from the water. This can be tested by applying a gas flame (lighter) underneath the panel for a few seconds. If the white shade disappears, it is most probably blushing. If not, it could be a water related residue on the surface of the lacquer itself.

(iv) **Adhesion**

Adhesion is naturally of great importance too, since containers are flexible. After all, one of their biggest advantages over glass is the fact that the metal container does not break after an impact, but only bends. The interior coating must adhere under the most stringent deformations before and after can making and filling. Coatings can only be flexible enough, if they adhere satisfactorily. It is not enough to test adhesion directly after curing the lacquer, but even more important to test it after deep drawing, sterilisation or pasteurisation and importantly, over curing. The methods mainly used for testing adhesion are given in Table 8:

TABLE 8 : SOME METHODS OF DETERMINING ADHESION:

Crosscut On The Sheet Or Flat Metal
Crosscut On Drawn Parts And Flanges
Crosscut On Impact Tested Parts
Crosscut After Sterilisation Or Pasteurisation

A good adhering adhesive tape should be used to remove non – adhering parts of the coating from the metal. This is of importance, because most interior coatings contain lubricants or waxes which might reduce the adhesion of the tape, giving misleading results. Adhesion has to be tested as quickly as possible after wet tests, and not more than 10 minutes after the removal of the test pieces from simulant solutions. Coatings tend to recover and a long delay would result in false readings and interpretations.

(v) **Flexibility**

It is important to test the flexibility of can coating systems, but the laboratory technician of a raw material supplier or a coating company seldom has the equipment similar to that of a manufacturing line. Flexibility testing has to be simulated.

(vi) **Deep Draw Tests**

Reasonably priced machines are available for flexibility testing during a slow deformation process. Some of them are able to draw cups in different sizes (one after each other), in different ratios of width and height. Table 9 gives some examples.

TABLE 9 : EXAMPLES OF POSSIBLE DEEP DRAW CYCLES

	Width	Height
First Cup	2	1
Second Cup (made from a first cup)	1	1
Third Cup (made from a second cup)	1	2

Different tools are available, and even square cups with different edge angles can be drawn. Care must be taken, however, to grease (for example with lard) the coated panel before drawing, to prevent damage to expensive tools. After deep drawing, the degree of destroyed film on the flanks of the cups can be determined easily and the cups can be used to determine other properties, such as chemical resistance. A standard system of known flexibility should always be tested for comparative purposes.

(vii) **Impact Test**

This test is used to evaluate the flexibility of coatings during a fast deformation. The coated test piece is positioned in such a way that a falling weight impacts on the reverse side of the panel. Due to the fact that can coating substrates are rather thin (often below 0.1 mm thickness), the maximum possible force which does not destroy the metal sheet has to be established in advance. Suitable impact testers are readily available. In the case of standard tinplate (of approximately 0.1 mm thickness), the maximum applicable force is 30 inch/lb. The same impact force is applied to the coated test piece and possible cracking of the coating is observed with the aid of a magnifying glass. In some cases it is very hard to detect fine cracks. To facilitate visual inspection, the impact tested piece can be dipped for a few minutes into a 5% solution of copper sulphate in water.

This is then wiped clean and, in the case of cracks, a reddish precipitate of metallic copper can be observed.

(viii) **Wedge Test**

This test is also used to determine flexibility during a fast deformation process. A coated test panel of 100×100 mm size is cut and a continuous wedge formed through the impact tester. For this purpose, the tester is fitted with a suitable tool system. The maximum available force, normally 160 or 180 inch / pound is applied. The test pieces are then totally submerged in a copper sulphate solution, whose strength depends on the base material, as follows;

- for aluminium   10% copper sulphate plus 5 % hydrochloric acid
- for tin plate   5% copper sulphate plus 1% acetic acid

The copper sulphate solution should be in contact with the panel for at least 5 minutes. After cleaning the test piece with water (care has to be taken, not to wipe off precipitates from the wedge itself) and drying it with the help of a piece of soft cloth, a clear adhesive tape is placed over the wedge and is removed in one, fast movement. The length of wedge which has not been affected is measured, and expressed in % of 100. The higher the value, the better the flexibility and the less porous the film. At least 3 tests should be done for every system tested and a sample of known resistance should always be tested as a reference.

(ix) **Porosity**

Cups produced by a cupping machine can also be used to test the porosity of an interior can coating system The cups must be given an interior coating so that the test pieces are placed in the cupping machine with the coated surface downwards. The cups should be drawn carefully, so that tools do not scratch the coated surface. The cups are filled as to two thirds with a 1–2% solution of copper sulphate and a current is applied between the copper sulphate solution and the exterior, uncoated surface of the cup. The current which penetrates the coating is measured and this (normally mA / cm^3) gives an idea of porosity. At least 3 tests should be done for every system tested and results compared with those for a master sample.

(x) **Crosslink Density**

There are a number of tests which determine the degree of crosslinking and thus give information about possible resistance.

(a) *Solvent Resistance*
This is the most commonly used method for testing the degree of cure. It is usually determined by a soaked cotton swab, which is rubbed to and fro, across the coated surface. The number of double rubs, before the solvent affects the coating, gives an indication of the degree of crosslinking. Suitable solvents are;

- acetone
- methyl ethyl ketone

- methyl isobutyl ketone

- xylene (less aggressive)

Methyl ethyl ketone is the standard solvent used.

(b) *Resistance Against Boiling Solvents*
This is a short test. Many samples can be tested in a relatively short period of time to determine degrees of crosslinking. A small strip of approximately 10mm width of the coated metal is cut from the sheet (or cut out of a coated, collapsible tube), and put into a test tube which is a third filled with a suitable (high boiling) solvent. A good choice is benzyl alcohol, which boils at approximately 160°C and is at the same time very aggressive. The test tube is carefully heated over a burner (this must be done in a ventilated, closed compartment) and the contents kept boiling for 5–10 minutes. After that, the strip is removed and cooled immediately with water. Depending upon the degree of film destruction, discolouration and loss of adhesion, one can roughly estimate the degree of crosslinking. A master sample of known resistance should always be tested to give a reference marker.

An alternative method is to place the coated metal sample into an extraction thimble (Sohxlet) and expose it to refluxing solvent for a period of time.

(c) *Resistance Against Dyes*
This is a very short test. A small strip (of approximately 10 mm width) of the coated metal is cut from the sheet and put into a test tube which is filled as to a third with a suitable dye solution, for example a 5% solution of methylene blue in a water / ethanol mixture.

The degree of colouration of the film again gives an idea of the degree of crosslinking. This test is very sensitive and useful in determining the degree of crosslinking for a great number of samples, including development products.

5.   Food Laws

All industrial food handling is regulated by law in all developed countries. Everything which comes into contact with food falls under these restrictions and laws and, of course, the interior coatings of cans containing food, beer, beverages, medicine, animal food or related products, do exactly that. Making things even more complicated is the fact that nearly every country has its own set of laws. In practise however, the FDA legislation, officially only in the United States of America, serves as an international standard. Furthermore, at least in Europe, there are strong signals coming from the Council of Europe, to simplify this complicated matter by introducing one set of recommendations for the whole common market. The work on standardisation for all EU members and others is well under way. Then, there will probably be only two major food laws used as unofficial standards world wide, FDA legislation (food and drug administration) and European law. Material used for the manufacture of coatings for food and other packaging will be strictly controlled and only specified and approved substances will be used.

New chemicals or monomers potentially intended for the manufacture of coatings for food packaging are subject to extensive toxicological testing before their acceptance can be considered. The food law regulators assess toxicological data obtained from a variety of tests specified. Some are the Ames tests for cell mutagenity or animal feed tests of 40 days or more. Tests might altogether take a number of years and might cost as much as 1–2 million pounds sterling. Only materials with strong prospects of acceptance will be seriously considered as candidates for evaluation and inclusion in rigidly controlled lists. Depending upon the amount of testing conducted and the resulting data, materials can be allocated an SML – specific migration limit (in the EU only). If a reduced dossier is submitted to the toxicologists advising the EU regulators, known as the Scientific committee for food (SCF), a substance is automatically treated as if it were harmful. A precautionary principle is applied and whilst this applies in Europe, a different set of criteria is used in the USA. This results in anomolies in the standards for coatings and their raw materials which must nonetheless be fully understood and rigorously observed.

Because of all this, food laws are a great challenge for researchers. Regardless of the final structure of the polymerised or polycondensed resin, they must only use raw materials listed in approved starting material lists. For the coating manufacturer, the raw material producer and the end user (in this case the cannery), the best way to prove that an individual coating is safe for use, is to test the material under appropriate conditions at an independent institute. This will give a good picture of migration with different simulants and of specific migration of, for example, critically viewed products like bisphenol A, bisphenol A diglycidyl ether (BADGE) and monomers such as phenol, cresol or formaldehyde.

6.   Suppliers Of Phenolic Resins

The basic chemistry to produce phenolic resins is, as said before, very old and well known. However, the number of manufacturers of phenolic resins is small, compared with the manufacturers of other synthetic resins. To a major extent this is because of the highly toxic nature of some of the raw materials used. It is also due to the potential exothermic reactions involved in the manufacture of resoles and the difficulties of treating the produced waste water. Table 10 gives the most important suppliers of phenolic resins for the coating industry in alphabetical order:

TABLE 10 : SOME SUPPLIERS OF PHENOLIC RESINS

ADVANCED RESINS South Glamorgan CF7 7PB Wales UK Brand names: Advorez
BAKELITE GESELLSCHAFT D-58642 Lethmathe Germany Brand names: Bakelite resin
CLARENCE RESINS Williamsville, N.Y. 14221 USA Brand names: Clar-Rez
DSM RESINS 8022 AW Zwolle The Netherlands Brand names: Synresene, Uravar
GEORGIA PACIFIC Deatur, Georgia 30035 USA Brand names: Bakelite phenolic resin
MONSANTO COMPANY 63166 St. Louis/MO USA Brand names: Santolink
OCCIDENTAL CHEMICAL B-3600 Genk Belgium Brand names: Occidental
RESINAS SYNTHETICAS S.A. E-08021 Barcelona Spain Brand names: Resibon
REICHHOLD CHEMIE AG A-1222 Vienna Austria Brand names: Beckacite, Super Beckacite, Varcum
SCHENECTADY de France F-62404 Béthune France Brand names: Schenectady
SHELL CHEMIE GmbH D-65760 Eschborn Germany Brand names: Epicure
VIANOVA RESINS GmbH D- 65203 Wiesbaden Germany Brand names: Phenodur, Alnovol, Alresen

References

(1) DIN 50021 ISO1456/3768 ASTM B 117-73

(2) DIN 50017

(3) Dr. E. Lattewitz, Farbenpost Höchst AG 7/1980

(4) Dr. E. Schwenk, 80 Years Synthetic Resins Höchst AG (1982), 5

Index for Part I